# Springer Series in Materials Science

Volume 212

The Springer Series in Materials Science covers the complete spectrum of materials physics, including fundamental principles, physical properties, materials theory and design. Recognizing the increasing importance of materials science in future device technologies, the book titles in this series reflect the state-of-the-art in understanding and controlling the structure and properties of all important classes of materials.

More information about this series at http://www.springer.com/series/856

Vladimir P. Sergienko · Sergey N. Bukharov

# Noise and Vibration in Friction Systems

Springer

Vladimir P. Sergienko
Sergey N. Bukharov
Department 3—Frictional Materials Science
V.A. Belyi Metal-Polymer Research
Institute of the National Academy
of Sciences of Belarus
Gomel
Belarus

ISSN 0933-033X ISSN 2196-2812 (electronic)
ISBN 978-3-319-11333-3 ISBN 978-3-319-11334-0 (eBook)
DOI 10.1007/978-3-319-11334-0

Library of Congress Control Number: 2014950676

Springer Cham Heidelberg New York Dordrecht London



Printed on acid-free paper

Springer is part of Springer Science+Business Media (www.springer.com)

# Preface

The present book analyzes the basic problems of oscillation processes and theoretical aspects of noise and vibration in friction systems. The book presents generalized information available in the literature data and investigation results of authors in vibroacoustics of friction joints, including car brakes and transmissions. The authors consider the main approaches to abatement of noise and vibration in nonstationary friction processes. Special attention is paid to materials science aspects, in particular, to advanced composite materials used to improve vibroacoustic characteristics of tribopairs.

The book is intended for researchers and technicians, students, and postgraduates specializing in mechanical engineering, maintenance of machines and transport means, production certification, problems of friction, and vibroacoustics.

Gomel, Belarus

Vladimir P. Sergienko
Sergey N. Bukharov

# Contents

# Acronyms and Notations

| | |
|---|---|
| ADC | Analog–digital converter |
| AFM | Atomic force microscopy |
| AS | Approximate spectrum |
| DFT | Discrete fourier transform |
| DTV | Disk thickness variation |
| EPSI | Electron pulse speckle interferometry |
| FFT | Fast fourier transform |
| FM | Friction material |
| FTV | Friction torque variation |
| FS | Friction-induced self-oscillations |
| LDV | Laser doppler vibrometer |
| LM | Lubricating material |
| MDOB | Multidisc oil-cooled brake |
| SEM | Scanning electron microscopy |
| VMC | Vibration-measuring converter |
| $A_c$ | Contour area |
| $A_r$ | Actual contact spot |
| $D$ | Index of damping capacity |
| $E_д$ | Dynamic elasticity modulus |
| $F$ | Friction force |
| $I_h$ | Linear wear rate |
| $L_p$ | Sound pressure level |
| $N$ | Normal to the friction surface force |
| $P$ | Pressure |
| $S$ | Area |
| $T$ | Temperature |
| $c$ | Viscosity |
| $f$ | Oscillation frequency (sound) |
| $f_p$ | Resonant frequency |
| $g$ | Free fall acceleration |
| $h_n$ | *n*th mode attenuation factor |

| | |
|---|---|
| $k$ | Stiffness |
| $m$ | Mass |
| $p$ | Sonic pressure |
| $t$ | Time |
| $\upsilon$ | Velocity of relative displacement |
| $x$, $\dot{x}$, $\ddot{x}$ | Coordinate, velocity, and acceleration, respectively |
| $\varepsilon$ | Linear deformation |
| $\eta$ | Loss factor |
| $\lambda$ | Wavelength, complex eigenvalue |
| $\mu$ | Friction coefficient |
| $\rho$ | Density of medium |
| $\omega$ | Angular oscillation frequency |

# Chapter 1
# Introduction

The leading research centers and companies of the world engaged in designing and manufacture of the modern high-tech goods are actively solving the problems connected with vibroacoustics in mechanical and dissipative systems, which include also tribological objects intended for various mobile, aerospace vehicles, complex domestic appliances and other aims [1]. Elevated interest to this sphere of investigations is related in part with the ecological aspect since generated by the mechanical systems noise is treated by the UN European Economic Commission as an important ecological parameter [2–4].

Abatement of undesirable noise and vibration generated by the rubbing and vibrating solid bodies is especially acute in transport due to the presence of the nonstationary friction joints in mobile vehicles.

The nonstationary friction joints are characterized by varying in time friction characteristics, load, velocity, temperature and properties of the rubbing materials. The friction process is considered to be nonstationary if at least one of above-named parameters significant for the friction contact is varying. The braking systems and friction clutches can be related to the most widespread joints of nonstationary friction. Their operation is often associated with elevated noise and vibration levels. These phenomena impair safety, reliability and serviceability of machines, worsening their quality and competitiveness.

Vibroacoustic activity of the friction units in brakes and friction clutches leads, from the one hand, to acoustic and vibration discomfort of the machine user and from the other to lowered durability of just as separate parts, so the integral machine. What is more, this problem is difficult to predict [5–7]. While the papers on this topic are abundant in quantity, the design and experimental methods promoting comfort and competitiveness of above products by meeting the existing norms and standards are lagging behind.

It is interesting that the expenses on the experimental and theoretical studies of noise and vibration have made up these years directly or obliquely about 50 % of the total budget of the basic companies engaged in developing friction materials (FM) and braking systems [8]. The problems with vibration and noise are decided



V.P. Sergienko and S.N. Bukharov, *Noise and Vibration in Friction Systems*,
Springer Series in Materials Science 212, DOI 10.1007/978-3-319-11334-0_1

**Table 1.1** Criteria for selecting friction pairs

| Estimated parameter | | Criterion | Unit | Grade points |
|---|---|---|---|---|
| Friction coefficient | Run-in | Deviation of $\mu$ from $\mu_{av}$ | – | 8 |
| | Mean level | Deviation of $\mu_{av}$ from $\mu_n$ | % | 7 |
| | Mean level stability | Difference between $\mu_{av}$ and $\mu_{min}$ or $\mu_{max}$ | % | 9 |
| | Heat-induced reduction | Deviation of $\mu$ from $\mu_{av}$ | % | 6 |
| | Sensitivity to pressure | Deviation of $\mu$ from $\mu_{av}$ | % | 7 |
| | Sensitivity to temperature | Deviation of $\mu$ from $\mu_{av}$ | % | 8 |
| | Sensitivity to velocity | Deviation of $\mu$ from $\mu_{av}$ | % | 7 |
| Noise | Bench tests | Noise level | dBA | 8 |
| | Ride tests | Noise level | dBA | 7 |
| Vibration | Vibration cold/hot | Vibration level | dB/1 m/s | 8 |
| | DTV-induced vibration | Disc thickness variation | μm | 9 |
| Wear | Lining | Lining wear | mm/GJ | 7 |
| | Disc | Disc wear | mm/GJ | 8 |
| Physical properties of lining material | Compressibility cold | Lining thickness variation | μm | 7 |
| | Compressibility hot | Lining thickness variation | μm | 5 |
| | Thermal conductivity | Heat conductivity factor | Wt/(m • K) | 4 |
| | Swelling | Variation in dimensions | μm | 6 |
| | Shear strength | Maximal force | N | 6 |
| Ecological safety of lining material | Lead | Content | mass% | 4 |
| | Cadmium | Content | mass% | 2 |
| | Carcinogenic fibers | Content | mass% | 7 |
| | Other | Content | mass% | 7 |
| Wear debris | Build-up on disc and lining surface | Subjective estimate | – | 8 |

very often by the end of the design stage or even after its termination when preparing for the production startup. This is connected with extra-costs on design debugging in attempting to abate noise and vibration. Therefore, taking into account the repair costs and related expenses, the manufacturers make efforts to create the brake systems with improved vibroacoustic characteristics already on the design stage.

There exist the procedures of preselecting frictional parts for such units as automotive brake blocks. Different criteria for selecting tribopairs for brakes are presented in Table 1.1. These criteria are used by the world-known car manufacturers in agreement with the national and international rules and legal documents like SAE, ISO, DIN, EN, JASO, Euro Spec and other. Their effect on the car quality is estimated using a ten-point scale [8]. It is evident from the table that noise and vibration are significant criteria for the car quality determination procedure.

The reduction of noise and vibration in the brake and transmission systems is attained mainly through varying structure of the joint, e.g., by refining geometry of the tribopair, or by involving new elements, like damping ones, or other.

It should be noted that very poor information is presented in scientific literature on the effect of composition and structure of the rubbing bodies on the noise and vibration level of the total system. It is very actual today to study the possible ways of abating vibroacoustic activity of tribojoints by optimizing structure and composition of friction materials based on the novel trends in tribology and materials science [9].

## References


1. N.K. Myshkin, M.I. Petrokovets, *Friction, Lubrication, and Wear. Physical Foundations and Technical Applications of Tribology* (Fizmatizdat, Moscow, 2007), p. 367
2. Regulations of EC UNO, Uniform Provisions Concerning the Approval of Motor Vehicles of Categories M, N and O with Regard to Their Braking, vol. 13 (09)
3. Regulations EC UNO, *Uniform Provisions Concerning the Approval of Motor Vehicles Having at Least Four Wheels with Regard to Their Noise Emission*, vol. 51 (02)
4. Tractors and Machines Agricultural Self-Propelled, General Safety Specification. State Standard GOST 12.2.019 (2006). Introduction 09.12.05, (Belarus Institute of Standardization and Certification, Minsk, 2005), p. 15
5. H. Abendroth, Worldwide Brake—Friction Material Testing Standards, Challenges, Trends. Proc. 7th Int. Symp. Yarofri, Friction products and materials, Yaroslavl, 9–11 Sept 2008, pp. 140–150
6. H. Abendroth, B. Wernitz, The integrated test concept: Dyno-vehicle, performance-noise, B. *SAE Paper*, 2000-01-2774, 2000
7. V. Vadari, M. Albright, D. Edgar, An introduction to brake noise engineering. *Sound and Vibration* [Electronic resource], (2006), http://www.roushind.com. Accessed: 15 Sept 2006
8. R. Mowka, Structured development process in stages of OE-projects involving with Western European car manufacturer. Proc. 5th Int. Symp. of Friction Products and Materials Yarofri, (Yaroslavl, 2003), pp. 228–232
9. Y. Pleskachevskii, V.P. Sergienko, Friction materials with polymeric matrix: promises in research, state of the art and market. Sci. Innov. **5**, 47–53 (2005)

# Chapter 2
# Oscillatory Processes and Vibration

Oscillations represent the most spread type of motion showing the property of recurrence, i.e., periodicity. Mechanical oscillations (vibrations) exist in all fields of technology, in everyday occurrence and nature. There is no in fact a domain where one or another kind of oscillations arises. The vibrations not envisaged at designing engineering objects are considered as undesirable and unsafe. It is important to understand the causes exciting these vibrations, their behavior and foresee their course in order to minimize or avert the consequences. This chapter sets forth general information on vibrating processes in mechanical systems, which include the ones operating with friction. The key notions and quantities are considered as well as the basics of the theory of nonlinear processes indispensible for understanding physical phenomena in frictional units.

## 2.1 General Information on Vibration in Mechanical Systems

A process during which some physical quantity(ies) peculiar for this process is/are subjected to transitions from increasing to decreasing alternated with inverse transitions from decreasing to increasing is called the *oscillation process*. This process is commonly time-dependent [1].

The oscillation process is characterized by iterations of these transitions, but in some cases, the transition of a physical quantity from increasing to decreasing or vice versa may take place only once. Such phenomena may occur, e.g., in impulse processes.

Very often, the oscillation process is observed in some physical quantity characterizing only some portion of the process and does not occur in the others. So, the oscillation process may be displayed or not depending on which part of the process is being studied, what equipment and kind of mathematical apparatus is used to interpret the experimental and theoretical evidences.

A physical system in which the oscillation process is feasible is called the *oscillatory system*. The systems of this type may be mechanical, acoustic, electrical,



V.P. Sergienko and S.N. Bukharov, *Noise and Vibration in Friction Systems*,
Springer Series in Materials Science 212, DOI 10.1007/978-3-319-11334-0_2

electromagnetic, electromechanical or other depending on the physical phenomena taking place in them.

Investigations of the processes and physical systems, representation and consideration of their properties and regularities as those of the oscillating nature whether they be acoustic, electrical or mechanical, is the subject of the theory of oscillations (vibration theory). In the text to follow we shall dwell upon the acoustic and mechanical vibratory systems.

Specific oscillation processes in mechanical systems, in particular, the elastic vibrations propagating through the structures and solid components (e.g., metal, ceramics, plastics, etc.) are called *vibrations*. The standard terminology on vibration says that the vibration represents a kind of motion of a point or a mechanical system during which scalar values for this point are subjected to vibrations [1]. Very often, however, the terms vibration and oscillations are mixed and employed as synonyms. The type of vibration of the in-plant origin that propagates through the design units is also termed as a structural noise.

Along with the longitudinal waves (the only type of the waveforms propagating in both gases and liquids) the solid media are able to transmit the transverse, surface, bending, and some other waveforms resulted from the restricted dimensions of the structural elements. The waveforms are characterized by their intrinsic velocity and other features (e.g., the velocity of the bending waves depends upon the frequency and size of the plate). When propagating, the waveforms can transform from one into another imparting thereby a complex character to the vibrations in a structure with expressed frequency and instability (in response to external factors) in time and space.

Vibrations are first initiated in the solid units, and excite then oscillations in the environment, e.g., air or liquid. Nevertheless, there may inverse occasions arise when a flow of gas or liquid induces intensive vibrations in the contacting objects. Vibration as an oscillating process can be classified from the standpoints of periodicity as periodical, almost-periodic and nonperiodic (oscillating quantity). The periodic vibration is observed when all oscillation values characterizing the process are repeated in the same sequence and time intervals $T$. The least $T$ value is called the oscillation (vibration) period.

The value $f = \frac{1}{T}$ expresses the *vibration frequency* measured in Hertz (oscillations per second). For the case of a rotary motion, it is convenient to use the *angular* (circular) frequency that is equal numerically to the number of cycles (oscillations, full revolutions) per $2\pi$ s time interval. The angular velocity is measured in radians per second.

The vibration process occurs very often simultaneously with such phenomena as linear accelerations, in which case the total effect presents a sum of instantaneous vibrations and linear acceleration values. The periodic vibration may be monoharmonic or polyharmonic.

The almost-periodic vibration includes the oscillations presented by a sum of harmonic oscillations having incommensurable frequencies. The almost-periodic vibration might have a final time interval after which the oscillating values are repeated. This interval is called an *almost period* [2].

### 2.1.1 Vibration Parameters

Physical quantities that characterize oscillatory processes by variations in time are called the oscillating quantities. They can be real or virtual, scalar or vector, determinate or random.

The value of the studied oscillating quantity $x$ in some instant $t$ is called an *instantaneous value* of the oscillating quantity

$$x = x(t).$$

In the case of vibration, each coordinate value indicating a selected point location is called a *vibratory displacement* (vibrodisplacement) denoted as $s(t)$.

The *mean modulus of the oscillating quantity* is found as the mean arithmetic or the mean integral of the absolute magnitude of the oscillating quantities in a given time interval $T$:

$$\bar{x} = \frac{1}{T} \int_{t_0}^{t_0+T} |x(t)| dt. \tag{2.1}$$

If we have $n$ discrete values $x$ of the oscillating quantity then the mean value of the modulus will be

$$\bar{x} = \frac{1}{n} \sum_{i=1}^{n} |x_i|. \tag{2.1a}$$

The *quadratic mean of the oscillating quantity* is found as a quadratic root of the mean arithmetic or the mean integral value of the oscillating quantity square within the considered time interval $T$:

$$\tilde{x} = \sqrt{\frac{1}{T} \int_{t_0}^{t_0+T} x^2(t) dt}. \tag{2.2}$$

If we have $n$ discrete values of $x_i$ of the oscillating quantity, then the mean quadratic value is found as

$$\tilde{x} = \sqrt{\frac{1}{n}\sum_{i=1}^{n} x_i^2}. \tag{2.2a}$$

Vibration is often estimated in engineering by the first and second derivatives of vibrodisplacement in time and is, correspondingly, termed as *vibration velocity*

$$v = \frac{ds(t)}{dt} = \dot{s}, \tag{2.3}$$

and *vibration acceleration*

$$a = \frac{d^2 s(t)}{dt^2} = \ddot{s}. \tag{2.4}$$

To facilitate comparison of the results the researchers often use in practice the logarithmic levels. The logarithmic level of vibration velocity $L_v$ in dB is found by the formula

$$L_v = 20\lg\left(\frac{v}{v_0}\right), \tag{2.5}$$

where $v$—estimated mean quadratic value of vibration velocity, m/s; $v_0$—initial (limiting) value of vibration velocity.

Vibration acceleration is expressed analogously in the logarithmic units $L_a$, dB, relative to a given threshold level $a_0$, m/s

$$L_a = 20\lg\left(\frac{a}{a_0}\right). \tag{2.6}$$

The threshold levels of both vibration velocity and acceleration are indicated for each concrete case. In contrast to the noise where the audibility threshold is taken constitutes zero decibels (see Chap. 3), the vibration decibels are counted from the conventional reference vibration velocity equal to $5 \times 10^{-8}$ m/s, and vibration acceleration of $3 \times 10^{-4}$ m/s$^2$. The threshold of vibration sensitivity makes up about 70 dB.

It should be noted that the Standard ISO 2631-1-1997 request to use vibration acceleration as the basic measured parameter. Even in the case of a low frequency region or a low level when it seems preferable to measure vibration velocity, the standard prescribes to convert the velocity into the vibration acceleration.

*Vibration (oscillatory) power N* is found from the product of operating force $F$ and vibration velocity $v$.

$$\mathrm{N} = \mathrm{F}v\cos\varphi, \tag{2.7}$$

where $\varphi$ is an angle between the force direction and the vibration velocity.

### 2.1.2 Description of Oscillating Process

The oscillating process is determined by variations of the oscillatory quantities in time, and their instantaneous values can be expressed by the following means:

(a) mathematical relations;
(b) numerical tabulated values;
(c) graphs or diagrams;

A combination of sequential instantaneous values of the oscillating quantity within a given time interval defines a temporal mode shape. As a spatial mode shape we understand a configuration of a set of points of a vibratory system exercising periodic vibrations in the moment when not all deviations of these points from their median position are equal to zero. The word "spatial" is usually omitted.

The mean quadratic value ratio to the mean one is called a shape factor.

$$K_{\Phi} = \frac{\tilde{x}}{\bar{x}}. \tag{2.8}$$

To characterize an oscillating process in a given moment one should introduce the notion of the *phase*. The phase is measured in degrees or radians. It can be preset by, e.g., an aggregate of instantaneous values of oscillatory quantities and their derivatives in time able to unambiguously define the oscillatory process.

To understand some intricate oscillatory processes it will be helpful to display $x$, $\frac{dx}{dt}$ on the phase plane. So, we lay the value of function $x(t)$ on the abscissa, and its first derivative $\frac{dx}{dt}$ expressing the same time moments on the ordinate axis.

By displacing over the phase plane sequentially and correspondingly to the oscillatory process phase, the image point forms a *phase trajectory*. For definiteness, the phase trajectory is fit with an arrow to show the direction of the states to follow. The regularities peculiar for the phase trajectories have brought us to important conclusions on the character of the oscillatory process under study.

The extreme values of the oscillatory quantity per given time interval are the largest and the least values, their difference makes up a peak-to-peak value. The absolute values of the extremes are called the *peak values* $x_p = |x_{\max}|$. The peak value ratio to a mean quadratic one is termed as a peak factor used to consider the vibration velocity or acceleration in terms of the criterion of the vibration load on elastic systems

$$K_a = \frac{x_p}{\tilde{x}}. \tag{2.9}$$

It should be noted that the functional time dependencies of the real physical quantities are diversified and rather complex. The oscillatory quantity character may change with time significantly depending on whether the very quantity or its derivatives in time is considered. We shall discuss some of most accustomed types of oscillatory processes by way of practical examples.

### 2.1.3 Harmonic Vibration

The periodic vibration is termed a harmonic or sinusoidal one when its instantaneous values are proportional to a sine or cosine of the linear time function, i.e.,

$$x = A\sin(\omega t + \varphi), \tag{2.10}$$

where $A$ is the amplitude; ω—circular frequency; φ—phase.

The harmonic vibration amplitude represents the largest absolute value the harmonic oscillatory quantity can reach. The amplitude equals to a half the harmonic vibration peak-to-peak value. The argument $\omega t + \varphi$ is called a phase angle, where $\varphi$ is the initial phase angle or the initial phase.

The harmonic vibration is characterized by vibration displacements $s(t)$, vibration velocity $v(t)$ and vibration acceleration $a(t)$, of the next form:

$$s(t) = A\sin \omega t, \tag{2.11}$$

$$v(t) = A\omega \cos \omega t, \tag{2.12}$$

$$a(t) = -\omega^2 A \sin \omega t. \tag{2.13}$$

It follows from (2.10)–(2.13) that vibrovelocity is shifted by a phase angle $\pi/2$, relative to the vibration displacement, while vibroacceleration—by angle $\pi$. This means that the image vector of the harmonic vibration velocity is ahead of the displacement vector by $\pi/2$, and the acceleration vector outstrips the displacement by $\pi$. So, the displacement and acceleration of the harmonic vibration are found in the antiphase.

For harmonic vibrations, the quantities expressed by relations (2.11)–(2.13) will be of the kind

$$\left.\begin{aligned} \bar{x} &= \frac{2}{\pi}A = 0,6366A \\ \tilde{x} &= \frac{A}{\sqrt{2}} = 0,7071A \\ K_{\Phi} &= \frac{\pi}{2\sqrt{2}} = 1{,}11 \\ K_a &= \sqrt{2} = 1{,}41 \end{aligned}\right\}. \tag{2.14}$$

Polyharmonic vibration means that a material point under study is oscillating in line with a number of harmonic laws simultaneously. The polyharmonic vibration can be analytically presented in the form of a sum of simpler harmonic oscillations.

There often occur oscillations of a point (body) formed from summation of a few harmonic oscillations whose frequencies are not interrelated through strict dependencies. Such oscillations cannot be related to the class of periodic ones since even slight frequency variations in one of harmonic components within a certain time interval may change drastically the complex vibration mode [3].

### 2.1.4 Nonstationary Determinate Vibration

We relate all types of impulse and aperiodic oscillations to the nonstationary vibration.

Impulse accelerations may result from the impacts, leaps or bursts or the like phenomena. The impact processes are manifold and rather complex. A simplest idealization of an impact is presented by collision of two masses via a cushioning spring (Fig. 2.1).

In practice, the colliding parts of units and aggregates are experiencing damped oscillations and repeated impacts under lower acceleration values. Impulse accelerations may last from a few tens of microseconds (rigid systems) till a few hundreds of microseconds (damped systems). Accelerations in the rigid systems may reach tens of thousands (of $g$ = 9.807 m/s).

In a simplest case the form of the impulse acceleration is defined as follows [4]. Let mass $M_1$ be in the state of rest before the impact, while mass $M_2$ be moving towards $M_1$ at velocity $v_0 = \frac{dx_1}{dt}$. As the spring touches the body $M_1$ it begins to compress, thus creating accelerations in both $M_1$ and $M_2$, i.e.,

$$M_1 \frac{d^2 x_1}{dt^2} - c(x_2 - x_1) = 0, \tag{2.15}$$

$$M_2 \frac{d^2 x_2}{dt^2} - c(x_1 - x_2) = 0, \tag{2.16}$$

where $x_1$ and $x_2$ are displacements of masses $M_1$ and $M_2$, correspondingly.

Let us differentiate (2.15) twice and substitute obtained in (2.5) and (2.16) accelerations $\ddot{x}_1$ and $\ddot{x}_2$

$$\frac{d^4 x_1}{dt^4} + \frac{c}{M_1}\left(1 + \frac{M_1}{M_2}\right)\frac{d^2 x_1}{dt^2} = 0. \tag{2.17}$$

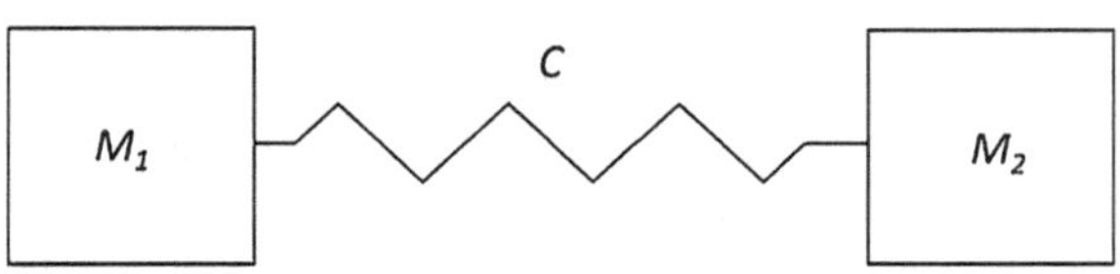


**Fig. 2.1** An idealized scheme of two masses impact

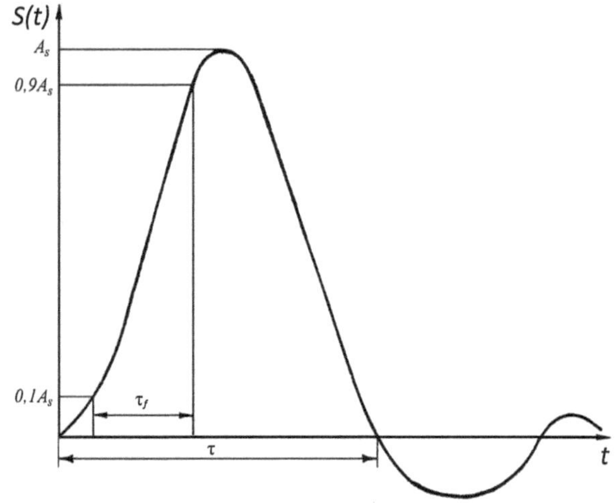


**Fig. 2.2** The impact pulse presentation in time domain [4]

So, (2.17) is solved in the form

$$\frac{d^2x_1}{dt^2} = a_1(t_0)\sqrt{\frac{cM_1M_2}{M_1 + M_2}}\sin\left(\frac{c}{M_1}\cdot\frac{M_1 + M_2}{M_2}\right)^{\frac{1}{2}}t. \tag{2.18}$$

From (2.18) we understand that the acceleration within the linearity limits presents a half-sine pulse.

The analysis of pulse (impact) processes admits application of just as time so frequency relations.

The characteristics of impacts in the time domain (Fig. 2.2) are the next: displacement dependencies $s(t)$, velocity $v(t)$ and accelerations $a(t)$; impulse amplitude ($A_s$, $A_v$, $A_a$); impulse duration $\tau$; impulse front duration $\tau_f$ [4].

To understand impact processes in the frequency domain we use Fourier integral to decompose the functions reflecting the impact as an aperiodic process with continuously varying frequency of the components from zero till infinity (see Sect. 2.3.1).

### 2.1.5 Random Vibration

A random process is a function of continuously varying argument $t$ presented by random values. The random vibration is characterized by a random behavior of variations in time of one or several parameters (amplitude, frequency, phase). Therefore, the results of a random vibration realization are not reproducible, i.e., they are random. In this connection, one should produce an ensemble of realizations to determine vibration parameters. A reliable description of random vibrations is fulfilled with the help of either probable or statistical characteristics (statistics).

In the case the statistics of the random vibration do not vary in time, the vibration can be called a *stationary* one. The vibration with changing in time probabilistic characteristics presents a *nonstationary* vibration. If statistic characteristics of the random vibration obtained by averaging in time coincide with the corresponding ensemble-averaged ones (ensemble of realizations), such a vibration process is called the *ergodic vibration.*

Random vibration $x(t)$ is described analytically by either integral $P(x)$ or differential function $p_x$ of distribution of the parameter being recorded or through the moment functions. A random value $x$ is fully determined by the probability distribution $P(x) = P\{X < x\}$, where $P$ is the probability of inequality $X < x$ existence. Random values $x_1, x_2, \ldots x_n$ are set by the $n$-dimensional distribution function.

$$P(x_1, x_2, \ldots x_n) = P\{X_1 < x_1, \ldots, X_n < x_n\}. \tag{2.19}$$

A random process $x(t)$ can be presented by a set of random coordinates $\{x(t_i)\}$ and preset by an integral distribution function of the kind

$$P(x_1, t_1, x_2, t_2, \ldots x_n, t_n) = P\{X(t_1) < x_1; X(t_2) < x_2; \ldots, X(t_n) < x_n\}. \tag{2.20}$$

Then, for a combination of random functions $\{x_1(t), x_2(t), \ldots, x_s(t)\}$ we determine an $n + s$-dimensional integral distribution function

$$\begin{aligned} &P(x_{11}t_1; \ldots x_{1n}t_n; \ldots x_{s1}t_1; \ldots; x_{sn}, t_n) \\ &\quad = P\{X_1(t) < x_{11}; \ldots X_1(t_n) < x_{1n}; \ldots; X_s(t_1) < x_{s1}; \ldots; X_s(t_n) < x_{sn}\}. \end{aligned} \tag{2.21}$$

The differential law of distribution of random process parameters is expressed by a derivative of the integral function of distribution

$$p(x_1, t_1; \ldots; x_n, t_n) = \frac{\partial n P(x_1, t_1; \ldots; x_n, t_n)}{\partial x_1, \ldots, \partial x_n}. \tag{2.22}$$

Along with the distribution functions, the analysis of the random process employs very often the moment functions. These functions are referred to as the mixed and simple moments that depend on the research target and are subdivided in their turn into the initial and central ones.

A mixed initial moment of the $k$-th order of a random function can be found from the relation:

$$\begin{aligned} m_{n1,n2,\ldots,ns}(t_1, \ldots, t_s) &= M\left\{[x(t_1)]^{n1}, \ldots, [x(t_s)]^{ns}\right\} \\ &= \int_{-\infty}^{+\infty} \cdots \int_{-\infty}^{+\infty} x_1^{n1} \ldots x_s^{ns} f(x_1, t_1; \ldots; x_s, t_s) dx_1, \ldots, dx_s, \end{aligned} \tag{2.23}$$

where $P(x_1, t_1; \ldots; x_s, t_s)$—$s$-dimensional probability distribution density; $M$—sign of averaging.

The mixed central moment of a random function is of the kind

$$d_{n1,n2,\ldots,ns}(t_1,\ldots,t_s) = M\left\{[x_0(t_1)]^{n1},\ldots,[x_0(t_s)]^{ns}\right\}$$
$$= \int_{-\infty}^{+\infty}\cdots\int_{-\infty}^{+\infty}[x_1 - m_1(t_1)]^{n1}\ldots[x_s - m_1(t_s)]^{ns}p(x_1,t_1;\ldots;x_s,t_s)dx_1,\ldots,dx_s. \tag{2.24}$$

Most applicable in practice are the following moment functions.

*The initial moment function of the first order*. Mathematical expectation of a random process

$$m_1(t) = M[x(t)] = \int_{-\infty}^{\infty} xp(x,t)dx. \tag{2.25}$$

For a stationary random process

$$m_1(t) = m_1 = \int_{-\infty}^{\infty} xp(x)dx. \tag{2.26}$$

The central moment of the first order equals to zero.

*Moment functions of the second order*. The initial moment functions of the second order are

$$m_2(t) = M[x(t)]^2. \tag{2.27}$$

The initial mixed moment functions of the second order

$$m_{1,1}(t_1,t_2) = M[x(t_1),x(t_2)]. \tag{2.28}$$

The central moment functions of the second order (variance)

$$d_2(t) = M[x(t) - m_1(t)]^2. \tag{2.29}$$

Correlation functions

$$R(t_1,t_2) = M\{[x(t_1) - m_1(t_1)][x(t_2) - m_1(t_2)]\}. \tag{2.30}$$

Notice that

$$R(t_1, t_2) = m_{1,1}(t_1, t_2) - m_1(t_1)m_1(t_2). \tag{2.31}$$

The correlation functions are time characteristics of the random vibration, i.e., they are determining the degree of correlation (statistical relation) between the random process values in different moments. There are two types of correlation functions, namely, the *autocorrelation* and *intercorrelation* functions. The autocorrelation function of a random process $x(t)$ in different moments of time $t_1$ and $t_2$ can be found from the relation:

$$\begin{aligned} R_x(t_1, t_2) &= M[x_0(t_1)x_0(t_2)] \\ &= \int\limits_{-\infty}^{+\infty}\int\limits_{-\infty}^{+\infty} [x_1 - m_1(t_1)][x_2 - m_1(t_2)]p(x_1, t_1, x_2, t_2)dx_1dx_2 \end{aligned} \tag{2.32}$$

The intercorrelation function of two random processes $x(t)$ and $y(t)$ is determined by the next equation

$$\begin{aligned} R_{xy}(t_1, t_2) &= M[x_0(t_1), y_0(t_2)] \\ &= \int\limits_{-\infty}^{+\infty}\int\limits_{-\infty}^{+\infty} [x - m_{x1}(t_1)][y - m_{y1}(t_2)]p(x, t_1, y, t_2)dxdy, \end{aligned} \tag{2.33}$$

where $m_{x1}(t_1) = M[x(t_1)]$, $m_{y1}(t_2) = M[y(t_2)]$.

## 2.2 Nonlinear Oscillations

### 2.2.1 Nonlinear Mechanical Systems

It is acknowledged that any dynamic system transforms its functions (the input function is aligned with the outlet one), therefore, each system is characterized by a certain operator called a system operator.

The operator is *linear* if the result of its action on any linear combination of prescribed (input) functions turns to be a linear combination of the results of its action on each separate function with the same coefficient. i.e., it obeys the principle of superposition.

For a *nonlinear* operator the principle of superposition does not work or is true only with some definite input functions and their coefficients. A system is called a *nonlinear* one on condition its operator is nonlinear.

The equations describing the behavior of linear systems are always linear. In the case there is at least one nonlinear equation, the system will be also nonlinear. The differential equations of motion of a nonlinear system may contain either nonlinear functions of coordinates and velocity or linear functions with time-dependent coefficients [5].

The nonlinear effects may be induced in practice by the following factors [6, 7]:

Nonlinearity of elastic characteristics of individual components or materials of the deformed part that do not obey Hook's law (e.g., rubber) or a structure of some elastic element (e.g., a conical wound spring with its lower coils pressed against the bearing surface, due to which the number of the working coils is gradually diminishing with deformation; the elements with the original interference; clearances in joints; arresters that restrict deformation, and etc.);

Nonlinearity of dissipative characteristics of deforming devices (hydraulic, pneumatic, frictional), movable or stationary frictional joints with internal friction in materials.

In a simplest case, nonlinearity in a mechanical system is attributed to nonlinear dependencies of the positional forces versus coordinates or resistance forces (e.g., frictional ones) versus velocity. These dependencies taken with the opposite signs are called force characteristics of the systems with one degree of freedom (e.g. kinetic characteristic of friction). The forces of a mixed type can be observed in more complex systems.

*The positional forces* are the ones that depend only on the position of the mechanical system, i.e., on its coordinates. In the case the increment in the positional force of a system with one degree of freedom is opposite to the deviation of the system from its equilibrium, the force is called a *restoring force*. This means that $F_p x > 0$, where $F_p$—ordinate of the force characteristic, $x$—deviation. When the force increments from the equilibrium towards deviation, the force is the repulsive one $F_p x < 0$.

We differentiate between the next positional forces: the forces of elasticity, gravity, buoyancy (of a body immersed in a liquid), and attraction (in a magnetic field). The derivative $dF_p/dx$ for the elasticity forces is called a *stiffness factor*. In the case this factor increases at $x > 0$ and reduces at $x < 0$, the force characteristic is *hardening*, or otherwise, it is a *softening* one. For one and the same system the force characteristic may be hardening under some $x$ values or softening under other.

Table 2.1 lists the examples of the mechanical systems with nonlinear positional forces and their force characteristics.

The forces that depend only upon the velocity of the mechanical system (if their power during motion of a system is not identically equal to zero, i.e., they are unorthogonal to motion direction) are termed as the resistance forces. They are subdivided into the next types:

friction forces in immovable units (mechanical systems with the elements exercising a periodic or constant relative motion);
friction forces in movable units (mechanical systems whose elements are nominally immovable relative to each other);
internal friction forces in materials of design parts;
resistance forces of environment (gas or liquid).

**Table 2.1** Mechanical systems with nonlinear positional forces

| Mechanical system | | Force characteristic |
|---|---|---|
| Description | Diagram | |
| Weight pressed by a spring against a plane | 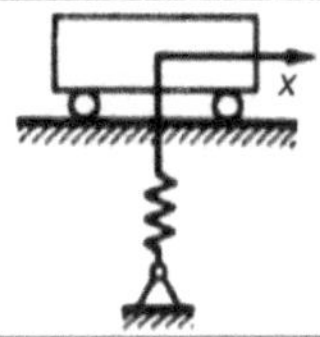  | 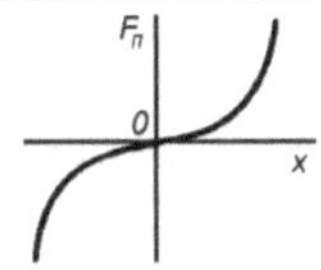  |
| Pendulum with immovable axis of suspension | 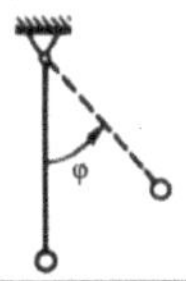  | 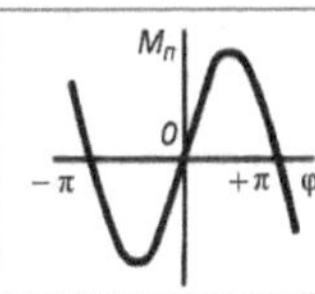  |
| Semi-cylinder with a longitudinal groove | 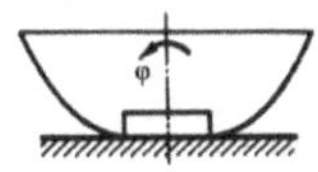  | 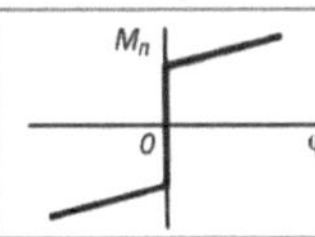  |

The resistance forces are on the main the nonlinear functions of velocity. This nonlinearity should be accounted for when determining the stationary self-vibration parameters and final amplitudes of oscillations at parametric resonance. This should be also done when studying transient processes in self-oscillating systems. The resistance forces most often encountered in practice are listed in Table 2.2.

The systems with one degree of freedom are usually based on the following simplified schematic representations:

$$\begin{aligned}&\text{force of linearly viscous friction } F(\dot{x}) = -k\dot{x},\\&\text{Coulomb's friction force } F(\dot{x}) = -k\,\text{sgn}\,\dot{x}\\&\text{nonlinearly viscous friction force } F(\dot{x}) = -k_1|\dot{x}|^{k_2}\,\text{sgn}\,\dot{x}, \text{ where } k, k_1, k_2 > 0,\end{aligned} \tag{2.34}$$

where the piece-wise regular function sgn $x$ is found as follows:

$$\text{sgn}\,x = \begin{cases} 1, & x > 0 \\ 0, & x = 0 \\ -1, & x < 0. \end{cases}$$

The resistance forces obeying inequality $F(\dot{x})\,\dot{x} > 0$ are performing a negative work resulting in dissipation of the mechanical energy, and are called *dissipative* forces. When $F(\dot{x})\,\dot{x} < 0$, the resistance forces are performing a positive work thus promoting energy gain in the system. These forces are called the *forces of negative resistance*. In the case the resistance force exercises intermittently a positive and negative work within different time intervals of motion, such a system may display self-oscillating properties.

**Table 2.2** Mechanical systems with nonlinear forces of resistance [8]

| Equations for resistance forces | Type of force characteristic |
|---|---|
| Exponential equation<br>$F(\dot{x}) = k_1|\dot{x}|^{k_2-1}\dot{x}$ | 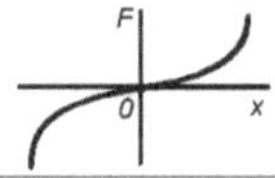 |
| Coulomb equation<br>$F(\dot{x}) = k\frac{x}{|\dot{x}|}$ | 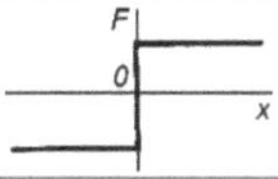 |
| Linear cubic equation<br>$F(\dot{x}) = k_1\dot{x} - k_2\dot{x}^3$ | 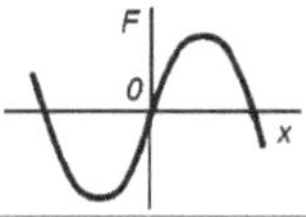 |
| Linear and Coulomb's equation<br>$F(\dot{x}) = k_1\frac{\dot{x}}{|\dot{x}|} + k_2\dot{x}$ | 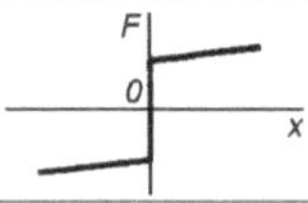 |
| Coulomb's, linear and cubic<br>$F(\dot{x}) = k_1\frac{x}{|\dot{x}|} - k_2\dot{x} - k_3\dot{x}^3$ | 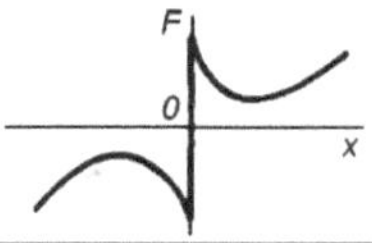 |

Self-oscillating properties impart also the forces of a mixed type to the system that depend on both coordinates and velocities. The forces that can be presented in the form of a product $F = F_n(x)F(\dot{x})$ are called the forces of positional friction. The examples of such forces are given in Table 2.3.

The friction force in the first system varies with the pressing force $F_t$ related with coordinate $x$; the pressing force $N$ in the second system remains invariable, although friction is observed at a rather high $F_t$, i.e., as soon as $x$ reaches some definite value. Characteristic for dissipative forces function $F(x, \dot{x})$ circumscribes during oscillations a hysteresis loop with the area equal to energy $W$ dissipated within a cycle (Fig. 2.3) [9]. The dissipative properties of a system at monoharmonic oscillations are conditioned by the hysteresis loop area and are independent of its form.

The group of nonlinear systems includes the inertialess linearities having insignificant delay. The *inertialess system* is a system in which the output function depends only on the input function in each given instant and is independent of the input function behavior till a given moment. The operator of the inertialess system is a common functional dependence between the input and output variables and presents a characteristic of this system. The inertial linearities are, correspondingly, the systems with delay and their functions are dependent not only on the input function value in a given moment but also on its variations till a given moment.

The characteristics of nonlinear systems can be subdivided into the weak nonlinearities that affect little the dynamics of a system in definite service conditions and essential nonlinearities that should be taken into account in dynamic calculations. The former include also the characteristics that could be substituted in the

**Table 2.3** Mechanical systems with nonlinear forces of positional friction

<table>
<tr><th colspan="2">Mechanical system</th><th rowspan="2">Force characteristic</th></tr>
<tr><th>Description</th><th>Diagram</th></tr>
<tr><td>Elastic piston entering frictionally a conical channel</td><td>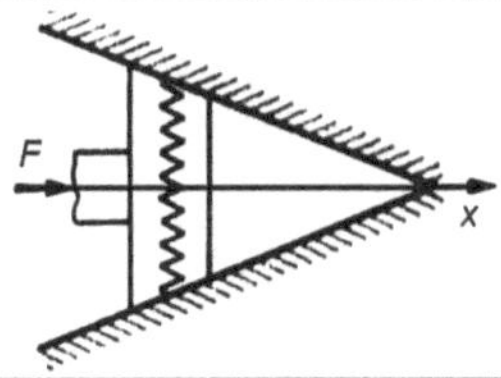
</td><td>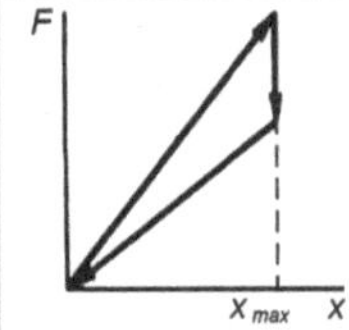
</td></tr>
<tr><td>Elastoplastic system with a slider</td><td>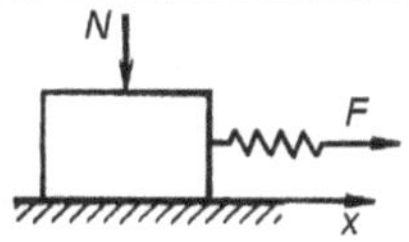
</td><td>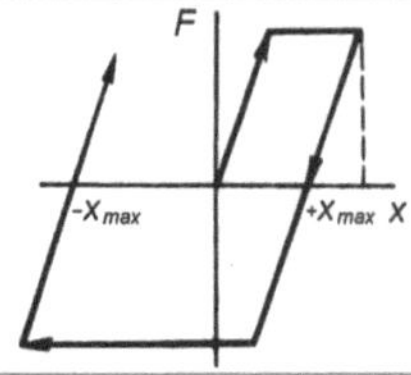
</td></tr>
</table>

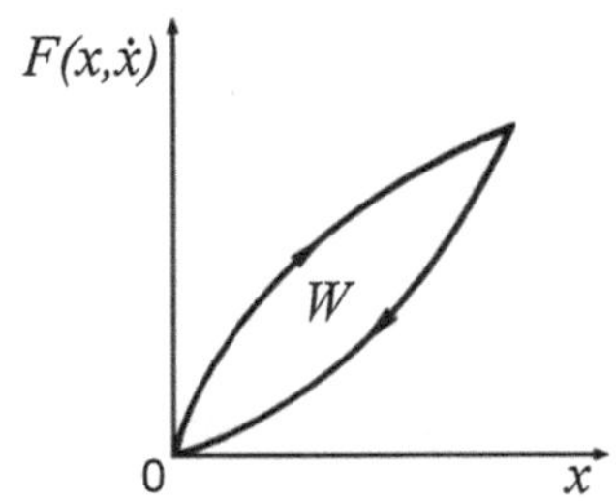


**Fig. 2.3** Dissipative characteristic of mechanical system

case of a narrow variation range of the input function or its negligible deviation from the mean altered value by the linear ones expressed through the unambiguous analytical functions or polynomials. For instance, a weak nonlinearity shown by the first example in Table 2.2, can be approximated by a low-power odd polynomial or a linear combination of sines. The essential characteristics include the essentially nonlinear functions, e.g. discontinuous or close to them functions (see example 2 in Table 2.2). The operator of these systems is usually presented by the piecewise-linear functions. In practice, the systems may display both types of nonlinearities.

In some cases, it is worthwhile using the approximate linear functions instead of the real nonlinearities, i.e., to exercise a *linearization*. However, one should at least approximately account for the mean nonlinearity of properties, which compels to preserve omission of the principle of superposition for the linearized systems.

Most simple for linearization is the case when nonlinearity of a characteristic is so small within the variation limits of the input function that it can be substituted roughly by a linear dependence determined by the first members of expansion of the characteristic in Taylor's series. Let the nonlinear characteristic $y = \psi(x)$ be an unambiguous differentiable function. If the input variable $x$ deviates but negligibly from some mean value $\bar{x}$, then we may use Taylor's formula, and by neglecting the remainders above the first order relative to $x - x_0$, we obtain

$$y \approx \psi(\bar{x}) + \psi'(\bar{x})(x - \bar{x}). \tag{2.35}$$

In terms of geometry, this linearization means substitution of the curve for a tangent to it in point $\bar{x}$.

The method of harmonic linearization (harmonic balance) developed by N.M. Krylov and N.N. Bogolyubov is used when the input function is of the form of sinusoidal oscillations of a constant amplitude $A$ and frequency $\omega$:

$$x = A \sin \omega t. \tag{2.36}$$

The output function is in this case also periodical, although nonharmonic and can be expanded in Fourier series:

$$y(t) = \psi_{\Pi} + \sum_{i=1}^{\infty} (A_i \sin i\omega t + B_i \cos i\omega t), \tag{2.37}$$

where $\psi_{\Pi}$ is the mean output variable value:

$$\psi_{\Pi} = \frac{1}{2\pi} \int_0^{2\pi} \psi(A \sin \varphi) d\varphi, \tag{2.38}$$

where $A_i$ и $B_i$—amplitudes of harmonic components, $i = 1,2,\ldots$:

$$\begin{aligned} A_i &= \frac{1}{\pi} \int_0^{2\pi} \psi(A \sin \varphi d\varphi) \sin i\varphi d\varphi; \\ B_i &= \frac{1}{\pi} \int_0^{2\pi} \psi(A \sin \varphi d\varphi) \cos i\varphi d\varphi. \end{aligned} \tag{2.39}$$

In the first approximation, the harmonics above the first one (overtones) are usually neglected:

$$y = \psi_{\Pi} + k_1 x + \frac{k_2}{\omega} \frac{dx}{dt}, \tag{2.40}$$

where the harmonic gain factors $k_1$ and $k_2$ are dependent on the linear system characteristic and the input variable amplitude:

$$k_1 = \frac{A_1}{A} = \frac{1}{\pi A}\int_0^{2\pi} \psi(A\sin\varphi)\sin\varphi d\varphi;$$
$$k_2 = \frac{B_1}{A} = \frac{1}{\pi A}\int_0^{2\pi} \psi(A\sin\varphi)\cos\varphi d\varphi. \tag{2.41}$$

To find a random input function we use the method of statistical linearization developed by I.E. Kazakov, R.K. Buton et al. It consists in the approximate substitution of the nonlinear characteristic $y = \psi(x)$ for an equivalent in the probabilistic sense dependence that is linear to a centered random input function:

$$y = \psi_n + k_c x_u, \tag{2.42}$$

where $\psi_n$ is a useful part of the output function. Notice, that its dependence on the useful part of the input function is a statistical characteristic of the system. In the case the odd characteristic of the system is $\psi_n = k_m x_M$, where $k_m$ is a statistical gain factor of the system in terms of the input function expectancy $x_M$; $k_c$—statistical gain factor of the system in respect to the random component: $k_c = \pm\frac{\sigma_y}{\sigma_x}$, where $\sigma_x$ and $\sigma_x$ are the mean quadratic deviations of the input and output variables, correspondingly; $x_u$—centered random component of the input function with a zero expectance.

The method of statistical linearization enables to define rather accurately the useful part of the output function and the level of fluctuations at the output, which implies determination of the transfer constant of the random component as a function, along with the useful and random components of the input functions.

In the case the input function of the system can be presented as a sum of the sinusoidal and random functions

$$x = x_M + A\sin\omega t + x_u, \tag{2.43}$$

we may use a combined harmonic and statistical linearization

$$y = \psi_n + k_{1c} A\sin\omega t + k_{2c} A\cos\omega t + k_{c\Gamma} x_u, \tag{2.44}$$

where the useful component $\psi_n$ in the case of the odd characteristic is taken proportional to a systematic component of the input function $\psi_n = k_{m\Gamma} x_M$, while the statistical gain factors $k_{m\Gamma}$ and $k_{c\Gamma}$ are presented by coefficients $k_m$ and $k_c$ averaged per variation period of the harmonic component; $k_{1c}$ and $k_{2c}$ are harmonic gain factors $k_1$ and $k_2$ for statistical characteristic $\psi_n$ obtained by statistical averaging. from function $\psi$.

Above-described combined linearization substitutes the nonlinear dependence of functions $y$ and $x$ for an approximate linear dependence between their main parameters (the constant or slowly varying components, amplitudes and phases of sinusoidal components and dispersed random components) and the approximate linear dependence between the quickly varying sinusoidal and random components.

## 2.2.2 Self-excited Vibration and Stability

Nonlinear mechanical systems are subdivided into the *autonomous* (off-line) in which the operating forces depend upon the condition of the system only (on coordinates and velocities) and *non-autonomous* (on-line) systems that include time in an explicit form into their differential equations of motion. The autonomous systems can be *conservative*, i.e., experiencing the effect of potential forces only, and *nonconservative* ones. In practice, we always deal with nonconservative systems in which the total energy is dissipated during motion.

There are two types of nonconservative systems, namely:

*Dissipative*, i.e., the systems experiencing the effect of dissipative and commonly restoring forces for which the equation of the energy balance is

$$\frac{dW}{dt} - F\dot{x} = 0, \tag{2.45}$$

where $W$ is the full energy; $F$—generalized force, being a function of coordinates and velocities. When the nonconservative force is interrelated with friction, it hampers motion: $F\dot{x} \leq 0$. The value of $W$ is always diminishing during motion. However, since energy cannot tend to $-\infty$, it approaches some constant value $W_0$ with time, while the product $F\dot{x}$ and, consequently, $\dot{x}$ tend to zero. This means that the system tends to a rest (equilibrium state). Only in the equilibrium, to which any system tends at whatever initial condition, the dissipative systems may acquire a stationary state, which is independent of the input function shift in time. They are unable to produce any stationary periodic displacements since the kinetic energy diminishes during motion;

*Self-oscillatory* systems in which periodic oscillations are probable. The loss of mechanical energy in such systems is immediately replenished by the energy from a source devoid of the inherent oscillatory properties. The energy supply from the source is controlled by motion of the system itself, while the period and peak-to-peak value are independent of the initial conditions within a wide range. Named vibrations are called stationary self-excited vibrations [10]. The process of gradual approach to stationary self-vibrations occurring after the arbitrary initial excitation of the system is called a *transient process*.

In contrast to free vibrations, self-vibrations are sustained and independent of the initial excitory impact. As compared to forced vibrations, the amplitude and frequency of the former are dependent rather on parameters of the system itself than on the external phenomena. However, this independence of the amplitude in respect to initial conditions is not always observed in its pure form. For instance, the constant amplitude of pendulum oscillations is interrelated with a wide range of initial conditions (when the initial deflection exceeds a certain value). Under some other initial conditions (when the starting deflection cedes this value) the oscillations are dying and the pendulum comes to a stop. Some self-vibrating systems may display several stationary processes with different amplitudes, where each process suits a definite relatively large range of initial conditions.

The incoming energy replenishes its inevitable losses in the system (or otherwise, the stationary periodic oscillations would not be probable) and simultaneously disturbs its stability. Therefore, it is typical for self-vibrating systems to loose stability in the equilibrium position in contrast to the dissipative ones. The energy flux is governed and converted most often via a nonlinear feedback of the vibrating system. The function of the nonlinear parameter is exercised in friction joints by the friction force.

If the nonlinear share of the force is rather small, the stationary vibrations are quasi-harmonic, and are described approximately by the next equation:

$$x = A\sin(\omega_0 t - \Phi). \tag{2.46}$$

The corresponding systems are called qyuasi-nonlinear. The differential equation of motion for such systems with one degree of freedom are of the form:

$$\ddot{x} + \omega_0^2 x = \psi(x, \dot{x}), \tag{2.47}$$

where $\omega_0$ is eigenfrequency of a corresponding degenerated system, $\psi(x, \dot{x})$—minor nonlinear function.

To solve (2.47) in the approximate form, we derive a function

$$\psi(x, \dot{x}) = \varphi(A\cos\varphi, -A\omega_0 \sin\varphi), \tag{2.48}$$

and come to a solution

$$x = A(t)\cos[\omega_0 t - \Phi(t)]. \tag{2.49}$$

The initial conditions are of the kind $A(0) = A_0$; $\Phi(0) = \Phi_0$. Function $A(t)$ is in the form of an envelope of the diagram of self-vibrations. With the unlimited time increase $t$, the amplitude tends to a limit in the form of a stationary self-vibration amplitude $A_{cm}$. This limit can be found from the condition of the amplitude constancy $\frac{dA}{dt} = 0$, which brings us to the equation

$$\int_0^{2\pi} \psi(A\cos\varphi; -A\omega_0 \sin\varphi)\sin\varphi d\varphi = 0; \tag{2.50}$$

Proceeding from this result, we can find $A_{cm}$.

A simplified solution of the nonlinear differential (2.63) was proposed by Van-der-Paul [11], where function $x(t)$ is presented in the form

$$x(t) = a(t)\cos\omega_0 t + b(t)\sin\omega_0 t \tag{2.51}$$

in provision that

$$\dot{a}\cos\omega_0 t + \dot{b}\sin\omega_0 t = 0. \tag{2.52}$$

Values $a$ and $b$ are constant for the linear systems, for the quasi-linear ones the time function is slowly varying.

Van-der-Paul's solution is true in the first approximation and leads to differential equations with separable variables:

$$\begin{aligned} \dot{A} &= \frac{1}{2\pi\omega_0}\int_0^{2\pi} \psi(A\cos\varphi; -A\omega_0\sin\varphi)\sin\varphi d\varphi; \\ \dot{\Phi} &= \frac{1}{2\pi\omega_0}\int_0^{2\pi} \psi(A\cos\varphi; -A\omega_0\sin\varphi)\sin\varphi d\varphi; \end{aligned} \tag{2.53}$$

where $\varphi = \omega_0 t - \Phi$.

Equation (2.47) can be also solved by the energy balance method. To simplify the solution, the true regularities inside every separate oscillation period are violated but their implementation within the whole period is observed, thus making the work per period equal to zero. The condition of the energy balance looks like

$$\Delta W = -A\omega_0 \int_0^{2\pi} \psi(A\cos\omega_0 t; -A\omega_0\sin\omega_0 t)\sin\omega_0 t dt, \tag{2.54}$$

where $\Delta W$ is energy increment of the system within a period per unit mass.

To estimate self-vibration parameters in the systems of solid bodies, one may apply linearization of the mathematical models of their natural vibrations [12]. In the case the force characteristic of the joints is described in terms of Coulomb's friction (Table 2.2), then

$$F(\dot{x}) = F_0\frac{x}{|\dot{x}|}, \tag{2.55}$$

where $F_0$ is a constant friction force, and linearization is made via substitution of Coulomb's friction for the equivalent viscous one. The equivalence factor $k_{э}$ is found from condition of equality of the friction force work per self-vibration period:

$$k = \frac{4F_0}{\pi\omega A}, \tag{2.56}$$

where $\omega$ and $A$ are estimated using the iteration method to solve matrix coefficients of a system of equations of motion.

Self-excited vibrations may differ much from the harmonic ones in conditions of essential nonlinearity of the system. They are called the *relaxation self-vibrations*. The examples of such vibrations are the Rayleigh and Van-der-Paul self-vibrating systems. Their equations of motion are, respectively:

$$\ddot{x} - k_1 \dot{x} + k_2 \dot{x}^3 + x = 0, \tag{2.57}$$

$$\ddot{x} - k_1 (1 - x^2) \dot{x} + x = 0. \tag{2.58}$$

In the second-order offline systems the phase space is presented as a plane. Motion of such systems is described either by a single differential equation of the second order:

$$\ddot{x} = \psi(x, \dot{x}) \tag{2.59}$$

or by a system of two differential equations of the first order:

$$\begin{cases} \dot{x} = y; \\ \dot{y} = \psi(x, y), \end{cases} \tag{2.60}$$

where $\psi$ is a known linear function or the output variable and its first derivative presenting phase coordinates of the system.

If we divide the second equation of system (2.60) by the first one, we shall obtain a differential equation for the phase trajectories

$$\frac{dy}{dx} = \frac{\psi(x, y)}{y} \tag{2.61}$$

which determines unambiguously the tangential to the phase trajectory in all points except for so-called special ones in which the following equalities are simultaneously met

$$\begin{cases} \psi(x, y) = 0; \\ y = 0. \end{cases} \tag{2.62}$$

Only one phase trajectory may pass through each point of the phase plane, except for the special points, which may initiate many trajectories.

In a general case, the phase plane of a nonlinear system may be rather intricate:

- the phase trajectories may behave differently in the vicinity of special points;
- special phase trajectories called separatrices may form borderlines between different parts of the phase plane;
- there may occur straight and curved switching lines corresponding to the angular points and break points of the polygonal or broken lines through which the piecewise-linear functions that substitute nonlinear ones are expressed.

There are nonlinear systems in engineering fit with isolated phase trajectories called *limiting cycles*. Every neighboring trajectory is either open or wound round a limiting cycle (i.e., the image point approaches it) or slides out of the cycle (the image point moves away). In the case the nearby phase trajectories are winding round a limiting cycle, it turns to be stable and the corresponding periodic motion of the system is stable. If the phase trajectories are withdrawing the limiting cycle, then the latter is unstable and the corresponding periodic motion of the system turns to be unstable too.

Let the friction force dependence versus velocity be expressed as follows (see Table 2.2):

$$F(\dot{x}) = k_1\dot{x} - k_2\dot{x}^3; k_1, k_2 > 0 \tag{2.63}$$

Small (curve 1) and grand initial perturbations (curve 2) are seen on the phase plane (Fig. 2.4), both belonging to transient processes. Curves 1 and 2 are approaching continuously a closed curve numbered 3, which is a stable limiting cycle.

Under small deviations from the equilibrium, the linear member of the friction force turns to be most important as a destabilizing factor. This makes equilibrium unstable and any arbitrary small initial perturbation may excite gradually growing vibrations, which, in their turn, promote a damping effect of the cubic member of (2.63), retarding thereby the vibrations and leading to a stationary self-vibration mode (Fig. 2.4b). At sufficiently large initial perturbations the damping effect of the cubic member gains more force than the destabilizing factor of the linear member,

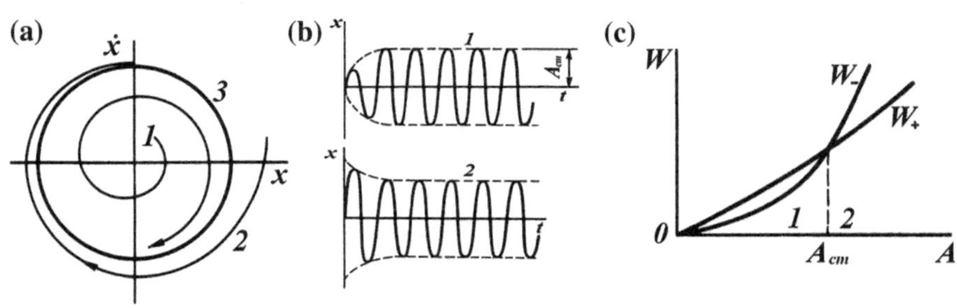


**Fig. 2.4** Self-vibrations in a mechanical system: **a**—phase diagram; **b**—coordinate variation in time, **c**—energy variation in the system at different vibration amplitude

wherefore the vibrations are damping initially. The influence of the cubic member attenuates with damping and the motion tends to the previous stationary mode (Fig. 2.4b).

Above-described two cases have been treated in work [8] from the standpoint of energy (Fig. 2.4), where $W_+$ is energy increment induced by the linear summand of the friction force; $W$; is the absolute energy variation value induced by the cubic summand. At low amplitudes $W_+ > W_-$ there occurs energy gain in the system, while with high amplitudes the energy outflows until the state $A_{st}$ realizes. Hence, a periodic mode is reached in the system with time in the case the system is out of equilibrium independently of the initial conditions. Such a state is called a self-excited vibration.

We differentiate between the following major features of nonlinear self-vibrating mechanical systems [6]:

- the possibility of a few equilibrium positions;
- free vibrations of conservative systems are non-isochronous, i.e., the frequency of free vibrations depends on their peak-to-peak value;
- the main vibrations are probable simultaneously with combined ones at the frequencies either larger (super-harmonic) or smaller (subharmonic vibrations, characteristic for, e.g. mechanical systems fit with an elastic arrester) by a whole number of times than the excitement frequency;

Self-excitement of vibrations can be either soft from the state of unstable equilibrium or rigid from the state of a stable equilibrium. The latter condition is illustrated on the phase plane in Fig. 2.5.

In the case there are several limiting cycles corresponding to a special point, the stable and unstable cycles will always be alternating. When an unstable focus is encircled by stable limiting cycles 1 and 3 with unstable limiting cycle 2 in between (Fig. 2.5a), then one of above-named self-vibration modes is established at any kind of perturbation. Such systems belong to the ones with a soft perturbation mode of vibrations. When a stable focus is surrounded by unstable limiting cycle 1 and a stable limiting cycle 2 (Fig. 2.5b), the mode of self-vibrations appears only under a rather strong perturbation at which the image point is found outside cycle 1. If the image point remains inside cycle 1, then vibrations are damping and such systems are referred to as the ones with a rigid mode of vibration excitement.

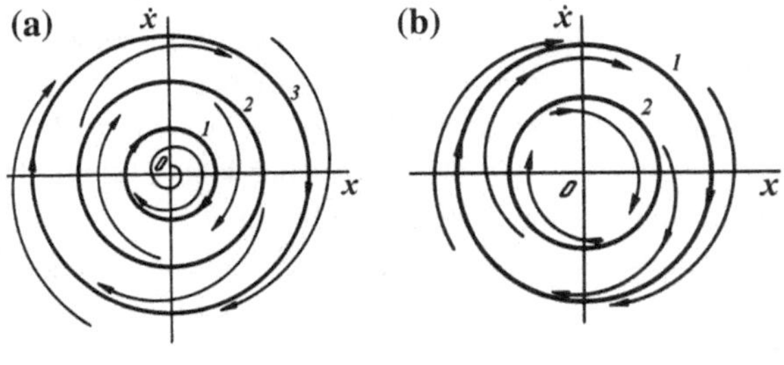


**Fig. 2.5** Phase diagrams of the systems with **a** soft and **b** rigid self-vibration excitement

The derivatives of the phase coordinates in special points of the system are equal to zero and are considered as those of the system equilibrium. Equilibrium stability in these points can be estimated using Lyapunov's method estimating the behavior of the system in the equilibrium vicinity.

Commonly, we do not know for sure some parameters of a mechanical system, or they can be occasionally altered with time. If the general properties of a system are varying but negligibly with slight variations in parameters bearing just a quantitative character, the system is termed as a structurally stable or a coarse one. In the case a small variation of some parameter leads to a qualitative change in the state of the system, it is called a structurally unstable or non-course system.

As far back as in 1892, A.M. Lyapunov has created a theory of stability that was true for any system that could be described by the differential equations. According to the theory, a system is considered to be stable if the perturbed motion deviation from the non-perturbed one under all $t > t_0$ is whatever small at any small enough initial perturbations in the moment $t = t_0$. The system is asymptotically stable if the perturbed motion deviation from the non-perturbed one tends to zero at $t \rightarrow \infty$. Since engineering considers only asymptotically stable systems to be virtually stable, so speaking about stability we shall further imply judt these systems.

Stability is as important for mechanical systems as the duration and behavior of transient processes. So, to estimate the quality of functioning proceeding from the transient process character, one may use time and damping decrement of the process, extremum values of the transient function and other parameters.

Let us bring the system of differential equations that describes the behavior of a mechanical system to a normal Caushy's form, i.e., to a system of the first-order equations solved relatively to the derivatives

$$\dot{x}_i = \psi(t, x_1, x_2, \ldots, x_n), \qquad i = 1, 2, \ldots, n, \tag{2.64}$$

where $x_1$, $x_2, \ldots$, $x_n$ are the quantities defining state of the system in a given moment.

If the right-hand sides of these equations are not explicitly dependent on time, then the system is not affected by time-dependent external perturbations and is, therefore, autonomous.

In contrast to linear systems, one and the same nonlinear system may be stable under some conditions and unstable under the other. In this connection, the notions of stability and instability are applied not to a very system but to its nonperturbed motion. As a nonperturbed motion is usually understood some desired motion or equilibrium condition of a system corresponding to its normal functioning subjected to perturbation-induced deviations, random ones including. The perturbed and nonperturbed motions are probable in one and the same system. They are defined by similar equations but at different initial conditions. If some probable motion with respective initial conditions $t_0$, $x_{10}^H$, $x_{20}^H, \ldots$, $x_{n0}^H$ and solution $x_1^H, x_2^H, \ldots, x_n^H$ is termed a nonperturbed one, then whatever the other motion be that differs from the chosen one, it will be called a perturbed motion.

So far, according to Lyapunov, stability can be formulated more strictly as follows:

A nonperturbed motion is called stable relative to quantities $x_1, x_2, \ldots, x_n$, if there exists a number $\xi > 0$, at which all perturbed motions obeying inequalities $\left|x_{i0} - x_{i0}^H\right| < \xi$ in any initial moment $t_0$ tend to a nonperturbed motion at $t \to \infty$: $x_i \to x_i^H$ at $t \to \infty$;
A nonperturbed motion is called unstable if there exists at least one number $\varepsilon > 0$ at which however small $\xi > 0$ be, the perturbed motion is observed that obeys condition $\left|x_{i0} - x_{i0}^H\right| < \xi$, under which inequalities $\left|x_i - x_i^H\right| < \varepsilon$ are not true for some $t > t_0$ values. In all conditions $i = 1, 2, \ldots, n$.

Above definitions are interpreted using the phase plane in Fig. 2.6. Its points correspond to a perturbed motion deviation from the nonperturbed one in each given time moment. The nonperturbed motion corresponds to a state of rest of the image point in the coordinate origin. The nonperturbed motion is stable in the case all phase trajectories beginning in the sphere of a small radius $\xi$ tend to coordinate origin at $t \to \infty$ (curve 1). It is unstable if the phase trajectories that begin close to the coordinate origin transcend at some $t > t_0$ the sphere of some radius $\varepsilon$ (curve 2).

To understand stability of nonlinear systems, A.M. Lyapunov has developed two general methods. The first of them is based on linearization of the equations describing behavior of systems. Usually, Taylor's linearization by expansion in series and truncation of the terms above the first power is made relative to deviation of the elements of perturbed motion in a system from the corresponding elements of nonperturbed one

$$\Delta x_i = x_i - x_i^H, \quad i = 1, 2, \ldots, n, \tag{2.65}$$

called as variations of variables $x_i$. The derived approximate linear equations we know as the equations in variations:

$$\Delta \dot{x}_i = \sum_{i=1}^{n} k_i \Delta x_i, \tag{2.66}$$

where coefficients $k_i$ are found by differentiation:

$$k_i = \frac{\partial}{\partial x_i^H} \psi_i(t, x_1^H, x_2^H, \ldots, x_n^H). \tag{2.67}$$

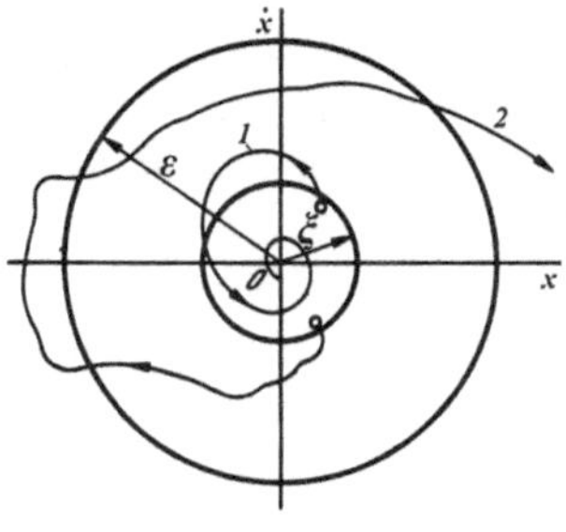


**Fig. 2.6** Phase diagrams of stable (*1*) and unstable (*2*) motions of mechanical system

In this case, the nonperturbed motion is *steady* when coefficients $k_i$ are constant. The research methods applied for stability of stationary linear systems can be used to study stability of a steady motion in nonlinear systems as well. In a general case, if $k_i$ is a function of time, it is worthwhile using the direct simulation methods with different initial conditions, as well as the coefficient "freezing" and other methods.

For the systems with significant nonlinearity we use the second Lyapunov's method implying a direct study of stability of a nonlinear system by finding such a function $\Pi(\Delta x_1, \Delta x_1, \ldots, \Delta x_n)$ of the coordinates of a phase space point of a given system that could be to some extent analogous to the potential energy of a material point resting in a common space. Further, similarly to the Legen-Dirichlet theorem, we admit that the points of the potential energy minima correspond to a stable equilibrium position, while those of the maxima are the positions of unstable equilibrium.

$\Pi(\Delta x_1, \Delta x_1, \ldots, \Delta x_n)$ is a constant-sign function if there is one and the same sign within a region containing the coordinate origin, except for some points where it equals to zero. The constant-sign function, equal to zero only in the coordinate origin is called the function of a fixed sign (fixed-positive or fixed-negative depending on the sign).

A nonperturbed motion is stable when the differential equations expressing the perturbed motion are such that a fixed-sign function $\Pi$ can be found. Its total derivative in time

$$\frac{d\Pi}{dt} = \sum_{i=1}^{n} \frac{d\Pi}{d\Delta\psi_i} \Delta\dot{x}_i \tag{2.68}$$

is a fixed-sign one that is opposite to function $\Pi$.

The second Lyapunov's method is restricted in view of difficulties encountered at determining $\Pi$ function. This concerns especially the systems for which linearization is inexpedient as it involves essential errors. That is why, the approximate methods are most often employed in engineering design. I.A. Vyshnegradskii was the first to propose the method based on the assumption that the properties of stability of a steady motion are displayed by a system even in the slightest perturbed motions generated within a short time interval after lending a minor initial perturbation to the system. Proceeding from this fact, all the terms above the first order relative to the coordinates and velocities are discarded and the conclusion on the nonperturbed motion stability is achieved in the form of integrals of the linearized equations. A combination of the research methods of stability based on linearized equations constitute the theory of the first approximation as opposed to Lyapunov's qualitative theory of stability.

We admit that in the first approximations the equations of perturbed motion present the first-order linear homogeneous differential equations with the constants:

$$\begin{aligned} \dot{x}_1 &= k_{11}x_1 + k_{12}x_2 + \cdots + k_{1n}x_n; \\ &\ldots \\ \dot{x}_n &= k_{n1}x_1 + k_{n2}x_2 + \cdots + k_{nn}x_n. \end{aligned} \tag{2.69}$$

These equations are the simplest when include only one variable. Such a form is called canonical

$$\dot{y}_i = k_i z_i, i = 1, 2, \ldots, n, \tag{2.70}$$

where $y_i$ are the coordinates connected linearly with $x_i$.

System (2.69) can be brought to (2.70) if the next determinant equals to zero

$$\begin{vmatrix} k_{11} - k & k_{21} & \ldots & k_{n1} \\ k_{12} & k_{22} - k & \ldots & k_{n2} \\ \ldots & \ldots & \ldots & \ldots \\ k_{1n} & k_{2n} & \ldots & k_{nn} - k \end{vmatrix} = 0 \tag{2.71}$$

Formula (2.71) is known as a characteristic equation. The nonperturbed motion of a linearized system and that of the initial nonlinear system is stable, provided all roots of the equation have a negative material part. It is unstable if at least one of the roots has a positive material part. When the material part of some of the roots equals to zero (in the case the material part of other roots is negative), then the motion of the linearized system is stable in provision that these roots have the corresponding simple elementary divisors. The motion is, vice versa, unstable when the roots correspond to multiple elementary divisors.

Determination of stability or instability is polynomial and corresponds to a developed determinant of the characteristic equation. This determination is probable without a predesign of determinant roots by using special Routh, Hurwitz, Nyquist, or other stability criteria based on acknowledged Cauchy theorem on the number of functional roots inside a closed loop. The frequency methods of analyzing stability, particularly, Nyquist's criterion, are commonly applicable for self-vibrations in mechanical systems.

## 2.3 Equipment for Vibration Measurement

Special-purpose instruments called vibrometers, are commonly used to estimate vibration parameters. They incorporate vibration-measuring transducers (VMT) as vibration receivers to transform vibration into electrical signals.

We differentiate between the VMT for measuring displacements, velocity and acceleration named, correspondingly, as vibrometers, velocimeters and accelerometers.

It should be noted, however, that more and more preference is given today to accelerometers due to the next reasons. Firstly, acceleration reaches significant enough values within the dominating medium and high vibration frequencies. For instance, vibration acceleration at vibratory displacement about 1 μm under 1,000 Hz frequency reaches 40 m/s$^2$ ≈ 4 g, where $g$ is a free fall acceleration. Secondly, the data on acceleration can be used to compute the dynamic inertial loads on design elements. What is more, modern vibration technologies offer a wide variety of VMT modifications to measure acceleration directly without any supplementary differentiation or integration.

Most popular among numerous VMT based on different physical phenomena (tensoresistive, potentiometric, piezoresistive, vortex-current, inductive, etc.) have turned to be piezoelectric devices (piezoaccelerometers) [13, 14]. The main advantages of these gages are their broad working frequency band, linearity of characteristics within a wide dynamic range, the output electric signal proportionality to the measured acceleration, high stability to external effects, exclusively high durability, technological effectiveness, possibility of operation without power sources, as well as a relatively small mass and compactness.

Piezoelectric accelerometers present the inertial transducers of the generator type for estimation of absolute acceleration. The sensitve element of the transducer consists of the inertial mass fixed in a case using an elastic element.

As soon as the object with a fixed VMT starts to oscillate, the piezoelement experiences the inertial load proportional to acceleration and mass of the sensitive element. According to a direct piezoeffect property (charge generation in response to mechanical load) a charge (voltage) is generated on the VMT contacts, which is proportional to acceleration. The acceleration transducers are fit with piezoelements operating under tension-compression, bending or shear. Named design peculiarities affect stiffness of the vibratory system, frequency of self-excited vibrations and the conversion factor. The accelerometer operating on the frequency below the resonance one shows in fact constant sensitivity. When the frequency exceeds the resonant one, its sensitivity rapidly falls (Fig. 2.7). The resonant frequencies of piezoaccelerometers are usually found between 10 and 100 kHz.

According to the State Standard GOST 30296-95 (IEC 1260, ISO 8041), the main technical characteristics of accelerometers are calibration, conversion factor, sensitivity, amplitude-frequency (AFC) and phase-frequency (PFC) characteristics, as well as frequency and temperature spans.

The calibration characteristic presents a dependence of the outlet voltage (of the charge) versus acceleration. This characteristic is linear, and the linearity factor does not surpass 1–5 %.

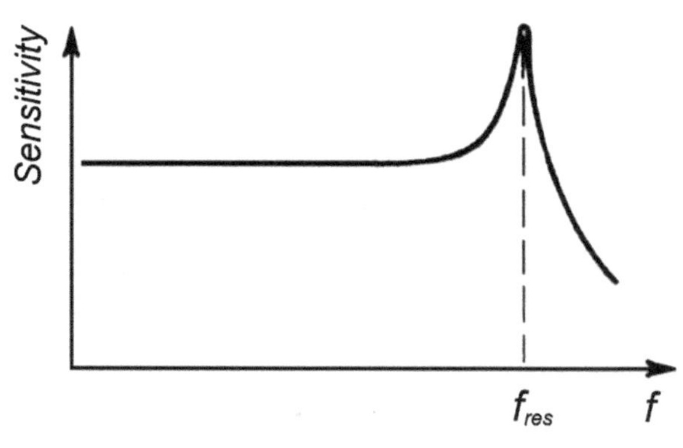


**Fig. 2.7** A typical frequency characteristic of piezoaccelerometer sensitivity

The conversion factor is equal to the outlet electric signal ratio to the acceleration value in the inlet to the VMT. It is used to determine the calibration slope.

It is important to ensure stiffness of the joint during installation or otherwise any fault may lead to the resonant frequency reduction and contraction of the working frequency band of the VMT. In this connection, a notion of a setting resonance was introduced that accounts for the eigenfrequency reduction (till 1.5–2 times reduction in practice). This factor depends upon the VNT mounting scheme on the object, namely, when it thrusts against a collar through the pad via a bridge.

The frequency range of piezoelectric VMT is an important characteristic. Its upper limit $f_\mathrm{u}$ depends upon the setting resonance $f_\mathrm{s}$. For majority of VMT $f_u = 0,3f_s$, while at nonuniform AFC it makes ≈10 %. The lower limit of the working frequency range depends upon the connecting cable capacity and the input resistance of the measurement instrument used. Rather high capacity (a few ths picofarads) and conversion factor are commonly shown by the VMT with the sensing element operating for bending.

One should bear in mind that piezoaccelerometers with a crosswise piezoeffect show sensitivity not only in the main direction (longitudinal) but in transverse directions as well. This may cause essential errors in analyzing complex vibration modes despite the fact that sensitivity in cross directions cedes the longitudinal one by as much as 20–30 dB. The transverse sensitivity can be reduced by adjusting symmetry of the mechanical system and the electric circuit, alignment of the gravity center of the movable mass with rigidity center, and by using several piezoelements able to average both mechanical and electric inhomogeneities of the VMT. Low values of the transverse piezoeffecrt factor of the order of 1 % are typical for the VMT which piezoelements are operating for bending and shear. The symmetrical VMT designs ensure hampering of the electromagnetic field and temperature effects.

The vibrometers consisting of three component fit with independent measurement channels of vibroacceleration in all three Cartesian axes are often used to study the complex vibration modes.

## References


1. Vibration. Terms and Definitions: State Standard GOST 24346-80 (Standart of Comecon 1926–79) Introduced 31.08.1980, Moscow, Izdatelstvo Standartov, 32 p. (1980)
2. G. Nor, *Almost Periodic Functions* (Gostekhizdat, Moscow, 1934), p. 130
3. A.E. Bozhko, *Reproduction of Vibrations* (Navukova Dumka, Kiev, 1975), p. 190
4. V.S. Pellinets, G.S. Skorik, *Modern Instruments for Impact Measurements* (BNIIKI, Moscow, 1973), p. 55
5. J. Stoker, Nonlinear Oscillations in Mechanical and Electrical Systems (Inostr. Lit., Moscow, 1956), p. 256
6. V.N. Chelomei (ed.), Vibration in Machinery, Refer Book in 6 Vols, vol. 2, ed. by I.I. Blekhman. Vibration of mechanical nonlinear systems (Mashinostroenie, Moscow, 1979), p. 351

7. I.I. Wolfson, M.Z. Kozlovskii, *Nonlinear Problems of Machine Dynamics* (Mashinostroenine, Leningrad, 1968), p. 284
8. Y.G. Panovko, Introduction in the Theory of Mechanical Vibrations (Nauka, Moscow, 1971), p. 240
9. V.L. Veits (ed.), Nonlinear Problems in Machine Dynamics and Durability (LGU Publisher, Leningrad, 1983), p. 336
10. V.A. Andronov, A.A. Vitt, S.E. Khiakin, *The Theory of Oscillations* (Fizmatizdat, Moscow, 1959), p. 915
11. B. Van-der-Paul, *Nonlinear Theory of Electrical Oscillations* (Gosizdat on communication techniques, Moscow, 1935), p. 42
12. Y.V. Demin, E.N. Kovtun, in *Estimation of Self-oscillation Parameters of the Systems with Coulomb's Friction*, ed. by V.F. Ushakov. Dynamic Characteristics of Mechanical Systems, Collection of Science Papers (Kiev, 1984), pp. 3–7
13. V.V. Klyuev (ed.), Instruments and Systems for Measuring Vibration, Noise and Impact, vol. 2. (Mashinostroenie, Moscow, 1978)
14. D.A. Grechinsky, V.N. Kovalsky, State of the Art and Promises in Development of Vibroacoustic Means. Instruments, Automation Devices and Control Systems. Review, *Instruments*, TSS-7 (2) (Moscow, 1988), p. 33
15. J. Hald, Combined NAH and beamforming using the same arra. Tech. Rev. **3**, 3–39 (2005, Bruel & Kjaer)

# Chapter 3
# Acoustic Radiation, Sound Waves and Fields

The present chapter presents definitions of the main acoustic notions and quantities along with the regularities and characteristics of the sound fields. Special instruments for measuring noise parameters of the studied objects are described. The features critical in measuring HF noise are stated. Information is presented on the methods of acoustic intensimetry and holography that are widely applicable in studying noise in friction systems.

## 3.1 General Quantities of Acoustic Radiation

*Sound* (noise) as a physical phenomenon corresponds to the waves of the elastic media. As a physiological phenomenon it characterizes sensing of *waves* by the ear.

Any disturbance of a 3-D point steady state in a solid, liquid or gaseous medium leads to perturbation in the form of waves propagating from this point. A 3-D region in which the perturbation occurs is called a *sound field*. The physical state of a medium in the sound field or, to be more exact, the changes in this state induced by the waves are commonly characterized by one of the following quantities:

(a) *sound pressure*, $p$ [N/m$^2$] presents the difference between the instantaneous value of the full pressure and the mean pressure observed in a medium without the sound field. The sound pressure is positive in the phase of compression, and negative in the phase of exhaustion;
(b) *vibration speed* of air particles, $v$ [m/s] presents an instantaneous value of a vibratory motion of particles during a sound wave propagation in the medium. The vibration sped is positive if the particles move in line with the sound wave direction, and it is negative if the particles are moving opposite to the sound wave propagation direction. Named values are the functions of time and coordinates.

The sound waves appearing in air are propagating from their driving point (sound source). Certain time is needed for the sound to travel from one point to



V.P. Sergienko and S.N. Bukharov, *Noise and Vibration in Friction Systems*, Springer Series in Materials Science 212, DOI 10.1007/978-3-319-11334-0_3 another. The speed of sound propagation depends upon characteristics of the medium and the mode of propagating in it sound waves.

The sonic speed in air at 20 °C temperature is 340 m/s. The sonic speed should not be mixed up with the vibration speed of air particles, $v$, which is a sign-varying value that depends on just as frequency, so sound pressure.

The sound wavelength, λ [m] measured along the sound wave propagation direction presents a distance between two neighboring points of the sound field in which vibration speeds of the particles in the medium are equal.

The wavelength in isotropic media is related to frequency $f$ and sonic speed through the formula

$$\lambda = \frac{f}{c}. \tag{3.1}$$

The sound waves are advancing together with the sonic energy.

The power transmitted per unit area in direction of the sound wave propagation is known as the sound intensity $I$ (Wt/m$^2$). The sound intensity is a time-averaged product of the sound pressure and the vibration speed of particles. In a general case, the sound intensity is described by the relation:

$$I = vp\cos(\theta) \tag{3.2}$$

where $v$ is a mean quadratic of the vibratory speed of particles in a sound wave, m/s; $p$ is a mean quadratic value of the sound pressure, N/m$^2$; θ is a phase shift between the vibratory speed and sound pressure.

If a sound wave propagates in a free sound field (in the absence of reflected sound waves), we obtain that

$$v = \frac{p}{\rho c}, \tag{3.3}$$

where ρ—density of the medium, kg/m$^3$; $c$—sound velocity in the medium, m/s.

The vibration speed and sound pressure in a free sound field are found within a phase, so far $\cos(\theta) = 1$. Consequently, the sound intensity in a free sound field in the wave propagation direction is expressed by the formula:

$$I = \frac{p^2}{\rho c}, \tag{3.4}$$

where ρc is a specific resistance of the medium to the sound.

The vibration speed measurements of the particles are often hard to perform. The vibration sped of particles can be described by a non-linearized Euler equation with a sound pressure gradient (i.e. variation rate of the sound pressure with distance). The *sound pressure gradient* can be measured by a ratemeter fit with two microphones placed close to each other (see a detailed description in Sect. 3.2.2).

Thus, the vibration speed of particles can be found by substitution of measurement results into Euler's equation.

The *sound energy density*, $w$, [J/m$^3$] is equal to the sound energy per unit volume. The sound energy density in a travelling plane wave can be found from the relation:

$$\varepsilon = \frac{I}{c} = \frac{p^2}{\rho c^2}. \tag{3.5}$$

The sound energy density is a scalar quantity that characterizes the sound field energy better than the intensity in the case the direction of the sound waves is indefinite, for instance, in enclosed spaces. The sound pressure and intensity characterize the sound field in some 3-D point. They are dependent on location of the sensor point, radiation direction and conditions of the sound wave propagation.

The *sound power of a source*, $P$ [Wt], presents a total quantity of the sound energy irradiated by a noise source into space in a unit time.

To find the sound power in a free sound field one should know the sound intensity, i.e., a mean sound energy flow per unit time referred to a unit surface normal to the sound propagation direction. If we sum up the sound intensity values in all directions generated from the source, we shall obtain the sound power

$$P = \int_s I_n ds, \tag{3.6}$$

where $I_n$ is the sound energy flow intensity normal to a surface element $ds$, Wt/m$^2$.

The direction of acoustic radiation is an important characteristic of any sound wave source (noise source). The real noise sources are usually characterized by a uniform radiation in different directions. Nonuniformity of sound radiation is described by a directivity factor

$$Q = \frac{p_n^2}{p_{ch}^2}, \tag{3.7}$$

where $p_n$—sound pressure measured at a certain distance from the source in a given direction; $P_{cp}$—sound pressure averaged over all possible directions at the same fixed distance.

The classification of sound fields usually makes allowance for the means and conditions of the sound wave advancement. Some sound fields and typical relations between the sound pressure and intensity are discussed hereinbelow. Notice, that these relations are accurately described in mathematical terms only in specific sound fields presented below, namely, in the free and diffusive fields.

### *3.1.1 Regularities of Sound Fields*

*Free Field.* The field in which sound waves are propagating in an idealized free space devoid of any type of reflection is called a free sound field. Such conditions are met in the open air (at enough distance from the ground) and anechoic (dead) chambers or spaces where the incident sound waves are fully absorbed by the walls. It is characteristic for the sound waves propagating in a free field to display a 6 dB reduction of the sound pressure level and sound intensity if to increase the distance (in direction of the sound wave propagation) from the sound source twice as much. Basically, this property obeys the inverse-square law. The sound pressure ratio to the sound intensity (more precisely, the ratio of their amplitudes) in a free sound field has been determined in mathematical terms. This mathematical relation enables to find the sound power radiated by a sound source in a free field (analogous method is described in [1].

*Diffusive Field.* The diffusive sound fields are characterized by a multiple reflection of the sound waves that leads to propagation of the waves in all directions with identical amplitudes and probability. The approximation of the diffusive sound field is presented by the fields enclosed in the reverberation chambers or rooms. Although the total sound intensity in the diffusive field equals to zero, there is a theoretical formula interrelating the sound pressure and a *one-sided sound intensity*. The one-sided sound intensity is the intensity in one direction in the case a similar constituent in the opposite direction is neglected. The one-sided sound intensity cannot be measured by a standard intensimeter (see Sect. 3.2.2), although it might be helpful together with the measured sound pressure in determining the sound power of a source in the diffusive sound field. The corresponding method is described elsewhere [2].

*Active and Reactive Sound Fields.* Propagation of sound waves is always connected with a flow of sound energy. However, the absence of the sound wave propagation does not necessarily exclude the presence of the sound pressure. A typical feature of the *active* sound field is a sound energy flow. The purely reactive sound field is, vice versa, devoid of the sound energy flow. The sound energy flow can be in any moment initiated by the sound source, but the radiated energy will obligatory return as soon as some time expires. The sound energy is accumulated analogously to accumulated in a spring mechanical energy. Consequently, the total sound intensity equals to zero. In general, all sound fields have both active and reactive components. The measurement results of the sound pressure in a reactive sound field may turn to be unreliable because the reactive component of the sound field is by no means connected with the sound power radiated by the source. Nevertheless, it is possible even in these conditions to measure the sound intensity with high enough reliability. Since the sound intensity is interrelated with the sound energy *flow*, the reactive constituent of the sound field does not generally affect the results of intensimetric measurements.

### 3.1.2 Use of Decibel Scale

The sound pressure, intensity and sound power values of a source are subjected to a wide-range variation. The sound pressure varies approximately within $2 \times 10^4$ till $2 \times 10^{-5}$ N/m$^2$, and their ratio makes up $10^9$.

For convenience, it is accepted in applied acoustics to estimate sound pressure, sound intensity, sound energy density and sound power in the relative logarithmic units called decibels instead of the absolute values.

Thus, we use a relative logarithmic scale instead of the corresponding scale of above-shown absolute values in order to abridge the range of readings. Each scale graduation expresses variations of these quantities not by a certain number of units but a certain number of times.

The *sound pressure level* $L_p$, dB, is found from the formula

$$L_p = 10\lg\frac{p^2}{p_0^2} = 20\lg\frac{p}{p_0}, \tag{3.8}$$

where $p_0$ is a threshold sound pressure equal to $2 \times 10^{-5}$ N/m$^2$. This value presents a threshold of audibility (zero level) at the sound level of 1,000 Hz.

Doubling of the sound pressure increases the sound pressure level by 6 dB, while its tenfold increase rises the sound pressure level by 20 dB.

The sound intensity level $L_I$, dB, is expressed by the relation

$$L_I = 10\lg\frac{I}{I_0}, \tag{3.9}$$

where $I_0$ is a threshold sound intensity equal to $10^{-12}$ Wt/m$^2$.

The logarithmic level units are not absolute but relative, and are, therefore, dimensionless. However, as soon as the threshold values of $p_0$ and $I_0$ have been standardized, the respective sound pressure and intensity levels have acquired the sense of absolute units since they define unambiguously the corresponding sound pressure and intensity values. Table 3.1 lists the data on the absolute sound pressure levels of different sources measured in a broad frequency range [3, 4].

Since the sound pressure level is dependent in a number of cases in the distance to the source, it should be always indicated.

The *sound power level*, $L_P$, dB, is found similarly to the sound pressure and intensity levels

$$L_P = 10\lg\frac{P}{P_0}. \tag{3.10}$$

Usually, $P_0 = 10^{-12}$ Wt is accepted as a threshold of the sound power.

**Table 3.1** Absolute sound pressure levels from different sources

| Source | Sound pressure level | Distance till source, m |
|---|---|---|
| Audibility threshold | 0–10 | – |
| Whisper | 30–40 | 1 |
| Low speech | 50–60 | 1 |
| Loud speech | 60–70 | 1 |
| Metal-cutting machine | 80–90 | 1 |
| Wood-working machine | 100–120 | 1 |
| Pneumatic tools | 110–120 | 1 |
| Piston engine | 120–130 | 3 |
| Jet engine | 130–140 | 3 |

*Radiation directivity index.* A non-uniform noise radiation from a source in different directions may be expressed not only by a radiation directivity factor but also using the directivity index.

$$DI = L - \bar{L}, \tag{3.11}$$

where $L$ is a sound pressure level measured in a given direction at a fixed distance from the source, dB; $\bar{L}$—sound pressure level averaged in all directions at the same distance, dB.

The directivity index DI is interrelated with the directivity factor Q through the formula

$$\boldsymbol{DI} = 10\,\mathrm{LgQ}. \tag{3.12}$$

### 3.1.3 Spectral Characteristics of Noise

A continuous succession of a combination of all frequencies within some time interval from the lowest $f_н$ till the highest boundary frequency. $f_в$ is called a *frequency range*. In the case the physical essence of the problem considered allows to subdivide the frequency range under study into sections, they will be called *subbands* or *frequency bands*.

The bandwidth is expressed in intervals. The interval that meets the condition $f_в = 2f_н$, is called an *octave*.

*The spectrum* is a key characteristic of noise as it indicates distribution of the acoustic energy within the frequency bands. It is taken into account in investigations, standardization, hygienic evaluation of noise, and etc.

The audible frequency range of 15–20 kHz is usually separated for spectral analysis into frequency bands to estimate sound pressure, intensity or power per each band. The noise spectrum is usually characterized by the levels of the quantities under study distributed over the octave frequency bands.

To investigate noise in more detail, one should use the octave frequency bands, for which $f_в = 2^{1/3} f_н = 1{,}26 f_н$.

The octave or the one-third octave band is commonly set by a center frequency $f_c = \sqrt{f_в f_н}$. Sometimes, the frequency range width is measured in percents relative to a medium frequency range.

There exists a standard series of the center frequencies of the octave bands within which we usually consider the noise spectra (from 31.5 Hz till 8.0 kHz) [5]. Depending on the frequency, we differentiate between the following noise modes:

low-frequency noise ($f_c < 250$ Hz);
medium-frequency noise (250 Hz $< f_c \leq$ 500 Hz);
high-frequency noise (500 Hz $< f_c \leq$ 8.0 kHz).

According to characteristics of its spectrum, noise is subdivided into the *wide-band* and *tone noise* [5, 6].

The wide-band noise is characterized by a continuous spectrum, and its width exceeds an octave.

The tone noise has a spectrum with expressed discrete (tone) components.

In practice, the tone noise is evaluated by making measurements in the one-third octave frequency bands. It does exist when the sound pressure level in one of the bands surpasses the neighboring ones by at least 10 dB.

### *3.1.4 Frequency Correction Scales*

Along with the linear decibel scale there exist many other scales that are applied in practice. Their level is also determined in decibels but within certain frequency bands and allowance for the frequency (or other) correction. For instance, scales *A*, *B* and *C* are used in instruments to measure noise (sound-level meters or noise meters).

Named scales play a significant role in the amplitude-frequency correction of the noise being recorded for subjective analysis of the sounds perceived by the man. The sounds of a similar intensity but different frequency are known to be heard by the human ear as different in volume. Figure 3.1 illustrates the curves of similar loudness. They indicate the level the sound of certain frequency should have in order to impose the same impression of loudness like the one produced at 1,000 Hz frequency with the level shown in the vertical column over the 1,000 Hz frequency value. Proceeding from the diagrams, the human ear is most sensitive in the

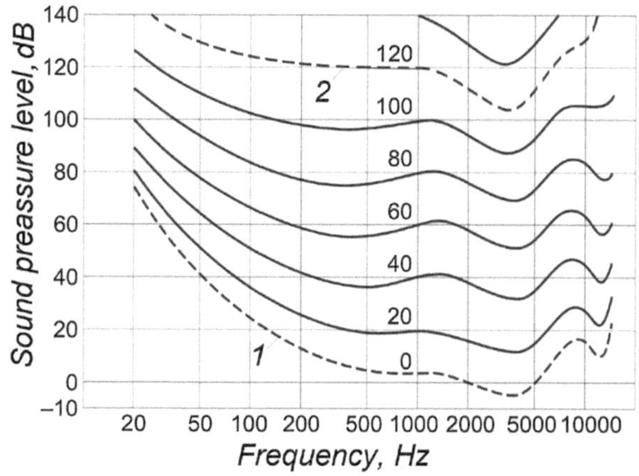


**Fig. 3.1** Curves of equal volume: *1*—audibility threshold, *2*—pain threshold [7]

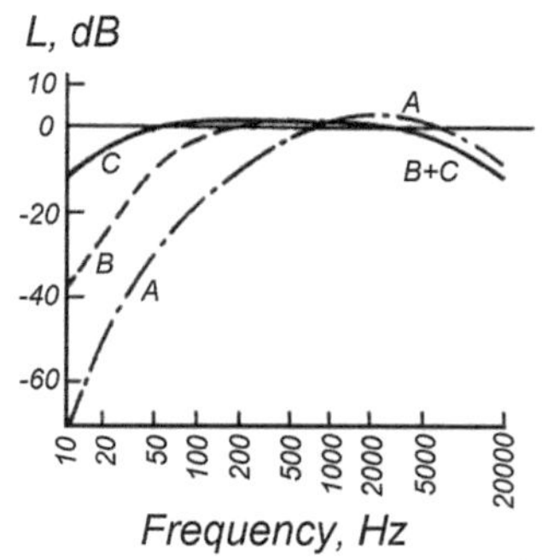


**Fig. 3.2** Characteristics of frequency correction *A*, *B* and *C* [6]

frequency range 3–5 kHz. The least audibility occurs within the low frequency range, but with increasing sound level the frequency characteristic of audibility sensitivity gains smoothness.

The curves of equal loudness were used in development of scales *A*, *B* and *C* with corresponding amplitude-frequency characteristics for sound-level meters: scale *A* for the low-level noise (0–55 dB), *B*—for the medium-level noise (55–85 dB), *C*—for high levels (above 85 dB). The frequency correction characteristics of the sound-level meters *A*, *B* and *C* (see Fig. 3.2) are in fact frequency characteristics of an average human auditory organ under various noise levels.

Scales *A*, *B* and *C* are used to make the integral estimates of the noise volume in all audible frequency ranges. However, the correction of frequency is not employed in the bounded octaves, one-third octaves and narrow (tone) bands. The measurement units in these scales are denoted, correspondingly as dB(*A*), dB(*B*) and dB(*C*) or dB*A*, dB*B* and dB*C*. In the recent years the scales *B* and *C* have practically come out of use since scale *A* has proved to perfectly fit the subjective perception of noise independently of its level. The sound pressure level according to scale *A* is presently called as a sound volume [7].

### 3.1.5 Time Characteristics of Noise

When interpreted in statistical means the quantity of noise can be *stationary* or *non-stationary*. The stationary noise is characterized by random processes having distribution functions independent of a zero-time reference. Most interesting from the practical viewpoint are the processes in which the first two moments of distribution are independent of time. Such noises are commonly called stationary in a broad sense. To the non-stationary ones belong the noises having at least one time-dependent statistical characteristic.

*Constant Noise*. A long-lasting noise is commonly generated by self-contained equipment operating uninterruptedly in one and the same mode, e.g., fans, pumps, computing devices, etc. Just a few minutes are needed in this case to measure the noise level using a portable meter. If it is possible to discern the tones and low frequencies, then the frequency spectrum can be measured and recorded, and further analyzed. The noise is considered to be constant if the sound pressure level alters by not more than 5dB*A* during measurements or a short-term process, e.g., during a working shift [5].

*Non-constant Noise*. The equipment operating in a cyclic mode, trains or cars passing by and flying past airplanes are generating a rapidly varying noise. The noise level of each machine cycle is measured by the method similar to the continuous noise but with account of a cycle time. When estimating the noise level each passing by car or train, or airplane is termed an "event". To determine the noise level of an event one should first measure the level of noise exposition (background noise) that unites noise levels and duration of the event in a single descriptor. In addition, it is recommended to use the maximal sound pressure level. An averaged value can be found from the noise level measurements in several analogous events.

Non-constant noise is commonly subdivided into the following types:
*fluctuating* in time noise, with a sound level varying continuously;
*discontinuous noise* with a stepwise varying sound level (by 5 and more dB*A*); an interval with a constant sound level lasts about 1 s or more;
*impulse noise* consisting of one or a few sound signals, each lasting les than a second. The noise arising from an impact or burst, e.g., at ramming, press blow, or gun shot is called the impulse noise. It consists of short shrilly noises with typical unexpectedness that irritates the man much stronger than can be expected if to judge only by the sound pressure level. To determine impulsivity of the noise we use the difference between the fast and slow reaction parameters. According to [8], this difference is to be not less than 7 dB. Besides, it is important to record the repetition frequency of the pulses (pulses per unit time).

A series of parameters analogous to the sound level has been developed based on the *A* scale. They are used to estimate individual localized in time noise events and noise modes during certain time intervals [6]. To compose such parameters we often take an *equivalent sound level* as a chief quantity. It is actually a sound level value in the *A* scale taken as a constant in time noise within the measurement duration with the mean square sound pressure value similar to the measured non-constant noise:

$$L_{Aeq} = 10\lg\left\{T^{-1}\int_0^T\left(\frac{p_A(t)}{p_0}\right)^2 dt\right\}, \tag{3.13}$$

where $L_{Aeq}$ is an equivalent (in energy terms) sound level of a non-constant noise, dB*A*; $p_A(t)$ is a current mean square sound pressure value of the noise measured with account of frequency correction in the *A* scale; *T*—preset time interval.

## 3.2 Methods and Equipment for Noise Measurements

### *3.2.1 Sound-Level Meters*

The sound-level meters represent the simplest traditional noise analyzers. They are commonly portable metering devices incorporating a measuring microphone, input amplifier, frequency filters that obey the standard parameters of the linear decibel scale and the frequency correction ones (*A*, *B* and *C*), an output amplifier and imaging means. A block-diagram of a typical sound-level meter is illustrated in Fig. 3.3 [4, 9].

Practically all sound-level meters are fit with the sockets for the external frequency filters (narrow-band, 1/3-octave, octave or other) to measure the spectral composition of the noise being measured. There is also a socket to take a measuring signal by an oscillograph or other gage. It is possible to choose the response speed of the sound-level meter according to the noise level variations, for which aim special rectifiers incorporate circuits with different time constants: *F—Fast*, *S—Slow*, *I—Impulse.* Figure 3.4 shows characteristic for the sound meter relative time dependencies recorded during interactions of a rectangular sound pulse.

Scale *I* is used to measure in fact any noise level, including impulse one, especially when the interest lies in the maximal sound level range to be found quickly. Scales *F* and *S* are intended to measure the noise devoid of the pulses, for which aim the mean indicator values are recorded. Some of the sound-level meters are equipped with a memory device enabling to memorize the maximal value of the noise observed during measurements.

Characteristics of the sound-level meters depend strongly on the microphone quality. Most often, they employ the microphones of a condenser, electret, or piezoelectric kinds. The latter are simpler and cheaper in contrast to the others. However, the condenser microphones ensure a high enough accuracy of

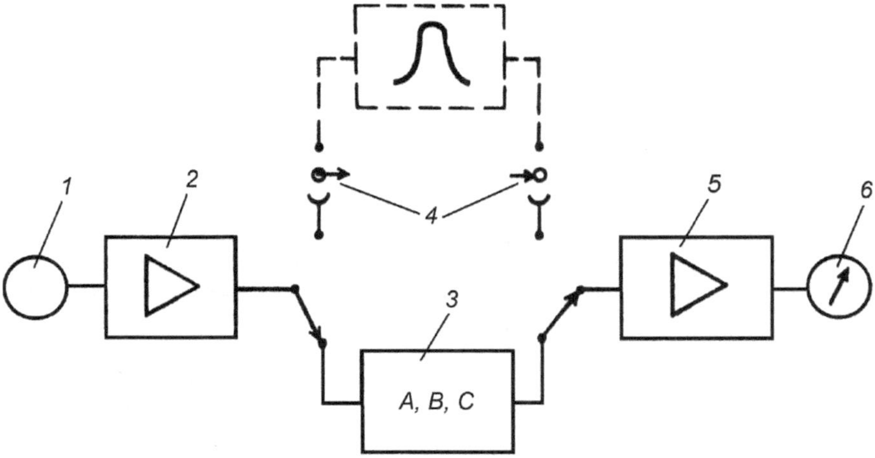


**Fig. 3.3** Block-diagram of sound-level meter: *1*—microphone; *2*—preamplifier; *3*—filters with standard frequency characteristics; *4*—connectors for external filters (shown by *dashed line*); *5*—output amplifier; *6*—indicator

measurements, wider frequency range (in the side of high frequencies) and better linearity of frequency characteristics. Depending on the measurement accuracy, the sound-level meters are subdivided into four classes, namely: zero class presents the meters for model measurements, class 1—accurate laboratory and in situ measurements; class 2—are used for normal accuracy and 3—for approximate measurements [8].

Strictly speaking, above-mentioned gages are intended to measure noise levels in the far field of the source. This is because their microphones are designed as the sound pressure transducers and the level measured is found according to formula (3.8). However, the levels obtained by this formula for the near-field values differ from those found by the main formula (3.9). In this connection, the sound-level meters used in the near-field measurements can estimate neither the sound field intensity nor the sound power of the source. The quantity measured in this case characterizes just the sound pressure level in the test point.

Another important characteristic to be considered in the measurements within the HF range (above 5–6 kHz for most of sound-level meters) is sensitivity to the microphone direction with increasing frequency. Because of this reason, the error may reach 3–5 dB in estimating noise level at 10–12.5 kHz frequency. In the case the noise sources are perfectly localized in space and the effect of the reflecting surfaces is insignificant, then the required corrections to the noise level values can be determined from a directivity pattern of the noise meter microphone. In the case the sound field under study is close to a diffuse one (incident waves in diffusive field are evenly spread in all directions), the corrections to the noise meter readings can be found in terms of the diagrams enclosed in specifications to the sound-level meters with allowance for the frequency A typical frequency dependence of the correction to be added to the gage readings is shown in Fig. 3.5.

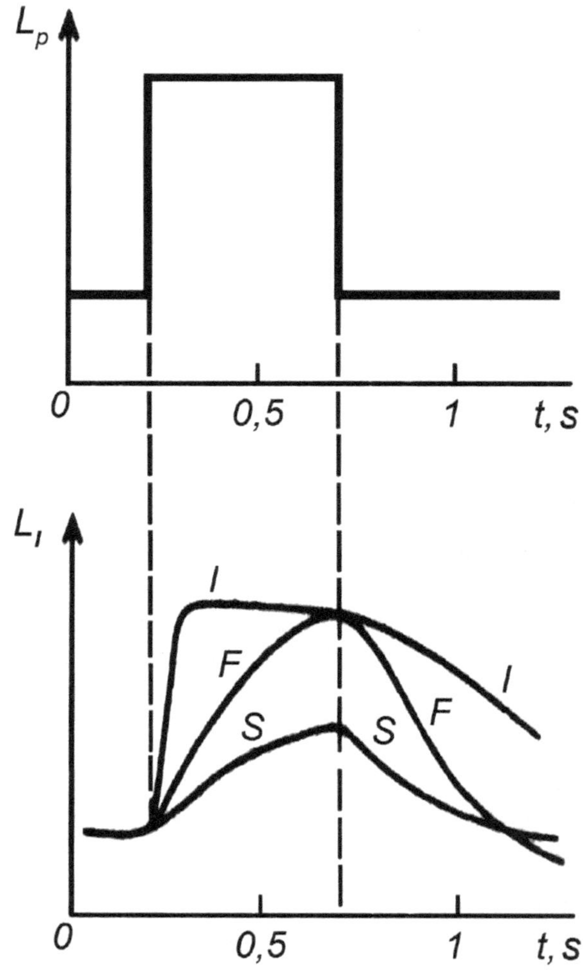


**Fig. 3.4** Time dependent readings of a sound-level meter: *F*—fast, *S*—slow, *I*—impulse

The sound-level meters are calibrated in the decibel scale relative to a standard sound pressure generated by a reference source (calibrator).

### *3.2.2 Acoustic Intensimetry*

Sound pressure measurements are not always giving comprehensive information on specific features of the sound field studied, especially the ones with a complex spatial structure. Profound is derived from the analysis of energy characteristics of the sound field, e.g., density of the potential and kinetic energy, intensity vector, etc. Investigations of energy behavior may in a number of cases help to understand structural peculiarities of the complex sound fields and regularities of their formation.

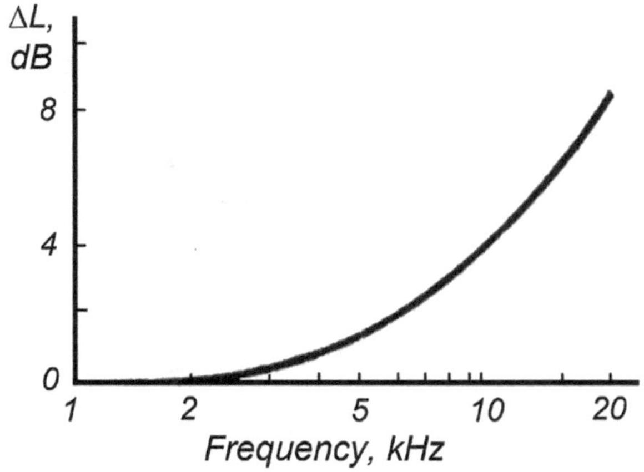


**Fig. 3.5** Frequency dependence of corrections ΔL of sound-level meter readings in diffusive field

It is also important to obtain information about the intensity vector. Determination of its magnitude and direction in various sound field points makes possible, in particular, to localize the noise source and calculate its acoustic power. Notice that the acoustic power and localization of the sound source can be determined by estimating its intensity in the near field. As it was underlined previously, the sound pressure measurements only in the near field may result in a faulty estimate.

Most applicable method of measuring sound intensity is today a so-called "two-microphone method". It includes two sound pressure receivers spaced at a much less distance than the wavelength is [10–15]. The sum of signals from the microphone of the acoustic intensimeter gives a mean value of the sound pressure in a point between the microphones (Fig. 3.6):

$$p = \frac{p(A) - p(B)}{2}, \tag{3.14}$$

where $p(A)$ and $p(B)$ are sound pressure values in the points where the microphones are located. It is assumed that pressure distribution in the space between the microphones can be accurately enough approximated by a linear dependence in view of smallness of this space as compared to the wavelength. To calculate the intensity, one should also know the vibrational speed. According to Euler's equation, this value is interrelated with the pressure gradient, while its component along the axis connecting the microphones (denoted as $x$-axis) can be roughly defined using two microphones proceeding from the finite-difference approximation. As a result, we have:

$$v_x = \frac{1}{\rho_0} \int \frac{p(B) - p(A)}{\Delta r} dt. \tag{3.15}$$

where $\Delta r$—space between microphones.

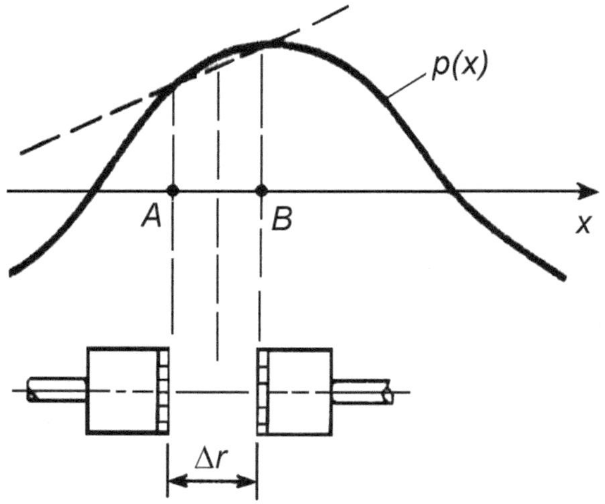


**Fig. 3.6** Determination of sound intensity by two-microphone method

It follows that the active intensity vector component along the axis connecting the microphones can be calculated by the formula

$$I_x = \overline{-\frac{p(A) + p(B)}{2\rho_0 \Delta r} \int [p(B) - p(A)]dt}, \tag{3.16}$$

The reactive intensity definition for inharmonic in time fields is derived from the Hilbert transform that shifts the phases of spectral components of the function being transformed (in our case, $v(\tau)$) per $\pi/2$:

$$J = \overline{\frac{1}{\pi} p(t) \int_{-\infty}^{+\infty} \frac{v(\tau)}{t - \tau} d\tau}, \tag{3.17}$$

where $\tau$—time delay.

Thus, taking into account (3.14), (3.16) and (3.17) the reactive intensity can be calculated as follows:

$$J_x = \overline{-\frac{1}{\pi} \frac{p(A) - p(B)}{2\rho_0 \Delta r} \int \left[ \int_{-\infty}^{+\infty} \frac{p(B) - p(A)}{t - \tau} d\tau \right] dt}. \tag{3.18}$$

Proceeding from above equations, a special hardware support is needed to determine the intensity by the two-microphone method. This equipment should include the units for obtaining the sum and difference of signals, integrators, multipliers, averaging blocks, and other facilities. Since there are difficulties in the hardware implementation of the intensimeters, the intensity measurements are often restricted to its active component.

Along with the processing algorithms based on the direct usage of equations (3.16) and (3.18) (called direct algorithms), we should also name widely applicable today algorithms that involve spectral notions [10–15].

The intensity vector component of a stationary noise field in a given direction $x$ can be expressed through a cross-correlation function $R_{pv_x}(\tau)$ of the sound pressure and the vibrational speed component in the same direction:

$$N_x = \overline{p(t)v_x(t)} = \overline{p(t)v_x(t+\tau)}|_{\tau=0} = R_{pv_x}(0). \tag{3.19}$$

The cross-correlation function $R_{pv_x}(\tau)$ is related to a cross-spectrum $S_{pv_x}(f)$ of the sound pressure and vibrational speed through Fourier transform

$$R_{pv_x}(\tau) = \int_{-\infty}^{+\infty} S_{pv_x}(f)e^{j\omega\tau}df. \tag{3.20}$$

From (3.19) and (3.20) follows that the intensity is related to the cross-spectrum through the formula

$$N_x = \int_{-\infty}^{\infty} S_{pv_x}(f)df, \tag{3.21}$$

i.e., the cross-spectrum is nothing less than a spectral density of the intensity. So, we can obtain for the active and reactive components of intensity

$$I_x = \int_{-\infty}^{\infty} \mathrm{Re}\left[S_{pv_x}(f)\right]df, \tag{3.22}$$

$$J_x = \int_{-\infty}^{\infty} \mathrm{Im}\left[S_{pv_x}(f)\right]df \tag{3.23}$$

If $F_p(A)$ and $F_p(B)$ are the sound pressure spectra in points $A$ and $B$ correspondingly, so approximation of spectra $F_p$ of the sound pressure and vibrational speed $F_{v_x}$ in a point between the microphones looks like:

$$F_p = \frac{F_p(A) + F_p(B)}{2}, \tag{3.24}$$

$$F_{v_x} = -\frac{F_p(B) - F_p(A)}{j\omega\rho_0\Delta r}, \tag{3.25}$$

After multiplying $F_p\ F_{v_x}^*$ and ensemble averaging, it is possible to obtain the cross-spectrum of the sound pressure and vibrational speed in the form

$$S_{pv_x} = \frac{j}{2\omega\rho_0\Delta r}(S_{AA} - S_{BB} + S_{BA} - S_{AB}), \quad (3.26)$$

where $S_{AA}$ and $S_{BB}$—autospectra, $S_{BA}$ and $S_{AB}$—cross-spectra of the sound pressures in points $A$ and $B$. Taking into account that $S_{BA} = S_{AB}^*$, we transform the difference $j(S_{BA} - S_{AB},)$ into $2\mathrm{Im}S_{AB}$. So, according to (3.22), the active component of the intensity is expressed via the cross-spectrum $S_{AB}$:

$$I_x = \frac{1}{\rho_0\Delta r}\int_{-\infty}^{\infty}\frac{\mathrm{Im}S_{AB}}{\omega}df. \quad (3.27)$$

For the reactive component we obtain from (3.23) and (3.26)

$$J_x = \frac{1}{2\rho_0\Delta r}\int_{-\infty}^{\infty}\frac{S_{AA} - S_{BB}}{\omega}df. \quad (3.28)$$

Proceeding from above relations, the active intensity component is defined by the imaginary part of the cross-spectrum, while the reactive one is found from the difference of the pressure autospectra in points $A$ and $B$. The transition to spectra $G_{AB}$, $G_{AA}$, $G_{BB}$ found in the frequency region (0, +∞) that can be obtained in the experiment brings us to a final result

$$I_x = \frac{1}{\rho_0\Delta r}\int_{0}^{\infty}\frac{\mathrm{Im}G_{AB}}{\omega}df. \quad (3.29)$$

$$J_x = \frac{1}{2\rho_0\Delta r}\int_{-\infty}^{\infty}\frac{G_{AA} - G_{BB}}{\omega}df. \quad (3.30)$$

The directional parameters of the two-microphone intensimeter depend on its property of measuring only the vibrational speed component that is directed along the microphone-connecting axis. Therefore, the directional pattern (in intensity) of above-described intensimeter is of a dipolar character $I_x = \left|\vec{I}\right|\cos\theta$, where $\theta$ is the angle between $x$ axis connecting the microphones and the intensity vector direction (Fig. 3.7). The phases of two directional lobes of the intensimeter are opposite to each other. This critical property of the intensimeter enables to localize the noise source.

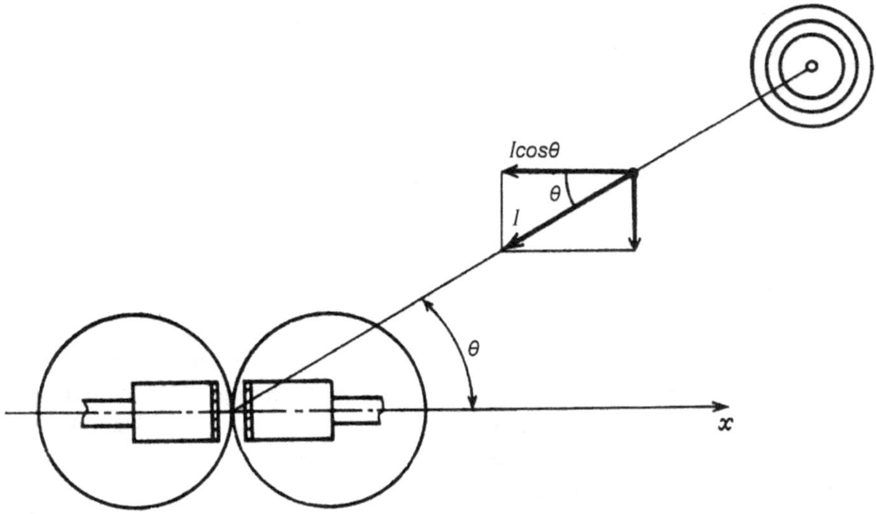


**Fig. 3.7** Directional characteristic of a two-microphone intensimeter

The two-microphone intensimeters impose certain limitations on the noise measurements. They are mainly connected with inaccuracies in approximations of (3.14) and (3.15) due to a finite spatial separation of the microphones $\Delta r$. Thus, the intensity measurements in the spatially inhomogeneous fields may show that the actual value of the intensity in the medial point between the microphones turns to be different from the measurement results. For instance, the measurement result $I_m$ in a spherical wave field is related to the real value of intensity $I_s$ as follows [11]:

$$\frac{I_s}{I_m} = \frac{\sin(k\Delta r)}{k\Delta r} \frac{1}{1 - \frac{1}{4}\left(\frac{\Delta r}{r}\right)^2} \tag{3.31}$$

where $r$ is the distance from the acoustic center of the source till the medial point between the microphones. It is evident from above relation, that with increasing $\Delta r/r$ value, i.e., as the intensimeter's sensor approaches the acoustic center of the source, the error in the measurements will augment. Although the restriction on the ratio $\Delta r/r$ is conditioned by the required measurement accuracy and is not at all connected with the studied field type (near or far field), it should be accounted for, especially when distance to the source is small, i.e., in the near-field region. It is indicated in [10] that a measurement error in the noise level arising from above-mentioned factor, makes up less than 1 dB, at $\Delta r/r > 1.1$ in the monopole field, at $\Delta r/r > 1.6$ in the dipole field, and at $\Delta r/r > 2.3$ in the quadrupolar field. It is evident that the restrictions on the distance between the source and intensimeter is not too tough in practice. What is more, there are the cases when the acoustic center of the source is often located inside the radiating surface.

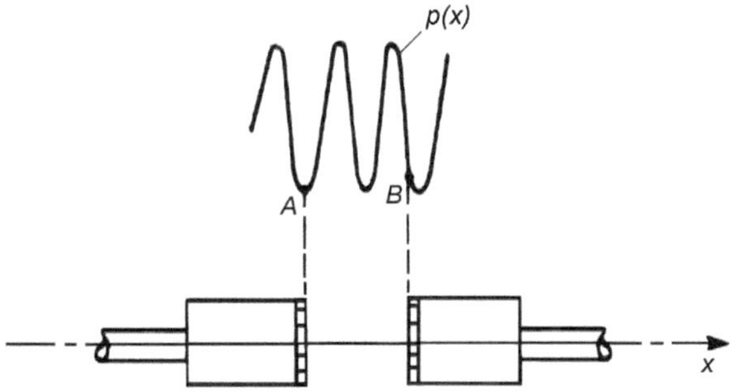


**Fig. 3.8** Two-microphone intensimeter in the HF sound field

Besides, it follows from (3.31) that the measurement error increases with increasing parameter $k\Delta r$. The reason is clearly seen from Fig. 3.8. If the frequency is so high that the spacing between the microphones becomes commensurable to the wavelength, then the approximations of (3.14) and (3.15) loose sense. So far, the upper limit of the useful frequency range of the intensimeter lowers with increasing distance $\Delta r$. From this point of view, $\Delta r$ value should be chosen as small as possible.

In practice, if to reduce distance between the microphones, the accuracy would impair within the low frequencies due to inevitable phase mismatch of two channels. Notice that under the effect of the sound wave of the same phase, the intensimeter channels generate the signals with somewhat different phases. As the frequency drops, the phase difference between the sound pressures in the points of the microphone location reduces and may become commensurable to a mismatch phase of the channels:

$$\frac{I_s}{I_m} = \frac{\sin(k\Delta r \pm \beta)}{k\Delta r} \tag{3.32}$$

where $\beta$ is a phase unbalance in the channels. The problem of measurement accuracy in the low the frequencies can be in part decided by correcting phase calibration of the intensimeter and exchange of the microphones [8, 12]. It is important to remember that the unbalance between the channel phases is also a reason of distortions in the directional characteristics of the intensimeter. Particularly, this may lead to a shift in a zero sensitivity direction by an angle $\psi = \arcsin[\beta/(k\Delta r)]$ like in the one shown in Fig. 3.9 [11].

So, the frequency range of the intensimeters depends upon the microphone spacing, which bounds its upper limit. It also depends upon their phase mismatch that defines the lower limit. The intensimeters having a large spacing between microphones are used in the Lf region, and with a small spacing—in the HF range. The relative sensitivity dependencies of intensimeters with different spacings between microphones (from 6 till 50 mm) are illustrated in Fig. 3.10 Notice that their LF part was drawn in supposition that the phase mismatch of the channels makes up 0.3° [11].

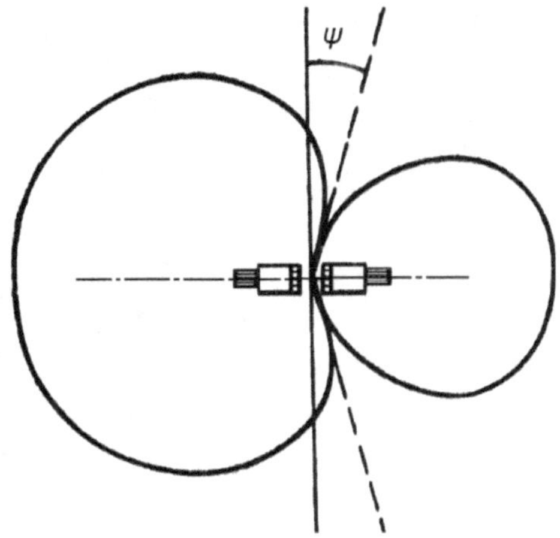


**Fig. 3.9** Distortion of directional characteristics of intensimeter due to phase imbalance in channels

### *3.2.3 Methods of Acoustic Holography*

Aside of above-considered single-component intensimeters (measuring only one spatial component of intensity vector) three-component intensimeters have been also developed that are equipped with six (three mutually perpendicular pairs) or four microphones [15] as well as different modifications of the flat and volume microphone gratings [16–18].

Such modern acoustic methods as the Beam-forming, Near-field Acoustic Holography—NAH, Statistically Optimal Near-field Acoustic Holography—SONAH are used to obtain intensity maps for the measured points along with the maps for close to the source planes and even the maps of the very surface of the source. This simplifies understanding of the results and promotes better resolution of the sources along with more accurate estimation of their qualitative characteristics.

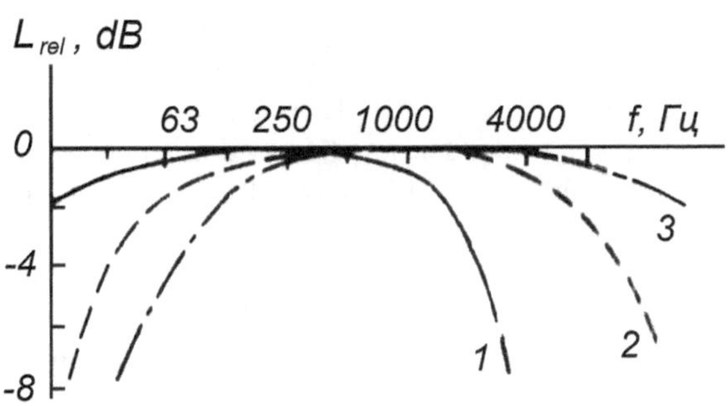


**Fig. 3.10** Relative measurement error by intensimeters with microphone spacing 50 mm (*1*), 12 mm (*2*), 6 mm (*3*)

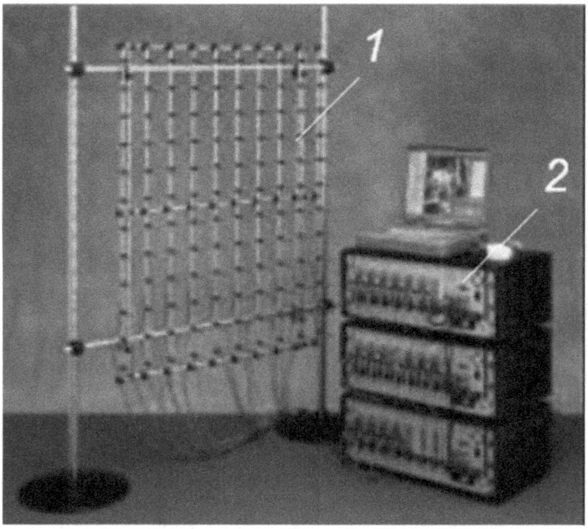


**Fig. 3.11** General view of a system for acoustic holography. The system includes antenna array (*1*) designed for 120 microphones and 132-channel (*2*) [19]

The formation of acoustic images (Beamforming) presupposes the construction of a map of the noise sources. For this aim, the noise levels are differentiated depending on the direction from which they originate. Named method can be used to construct the maps for a distant noise. It is especially useful in drawing maps for large objects.

NAH is based on measuring sound pressure in the near field of the source using flat microphone gratings and a multichannel analyzing system that performs fast Fourier transform of each microphone signal (Fig. 3.11). The essence of the NAH method consists in construction of a mathematical model to describe a sound field. The mathematical model is based on a set of measurement results of the sound pressure initiated as a rule not far from the source plane. Using this model it is possible to determine the parameters of the sound field, namely, pressure, intensity, acoustic speed of the particles, and so on, in the required planes parallel to the measurement one.

The transient process parameters are measured as a rule using large fixed antenna arrays in order to ensure a simultaneous measurement in all chosen points. This is commonly made in the form of intensity maps measured by the sensors located in some points directed towards the source. The methods based on the data recorded from the antenna arrays are intended to perfect named process since the data are recorded simultaneously from several points and to make the measurement much faster.

The acoustic waves generated by the working mechanisms or equipment and propagating in air are called the in-plane noise. Similar vibrations appearing as a result of operating car joints, trains, and so on are called the transport noise. The notion of "noise" does not impose any limitations on the characteristics of the sound or its spectrum.

It is important to emphasize that noise and vibration are in fact always generated due to one and the same reason, are running hand in hand, and are in some way interrelated. This is why, noise and vibration are commonly analyzed jointly. It is natural to handle experimental data by using identical digital signal processing methods, and to combat these phenomena by the same means that are able to abate both noise and vibration with a similar efficiency.

## References


1. Acoustics. Determination of sound power levels of noise sources by the sound pressure. Exact methods for anechoic and semi-dead chambers. State standard GOST 31273-2003 (ISO 3745:2003), 2005, p. 31
2. Noise of machines. Determination of sound power levels by the sound pressure. Exact methods for reverberation chambers. State Standard GOST 31274-2004 (ISO 3741:1999) (Standartinform, Moscow, 2005), p. 26
3. B.G. Prutkov, I.A. Shishkin, G.L. Osipov, I.L. Karagalina, *Sound-Proofing In Civil Engineering* (Stroyizdat, Moscow, 1966), p. 114
4. G.L Osipov et al., *Measurement of Noise Generated by Machines And Equipment* (Standard Publishing, Moscow, 1968), p. 147
5. Noise. General safety requirements. State Standard GOST 12.1.003-83 (Standard Publishing, Moscow, 1991) p. 14
6. J.D. Webb (ed.), *Noise Control in Industry* (Halsted Press, New York, 1976), p. 421
7. E.Ya. Yudin, *Noise abatement in industry. Reference Book* (Mashinostroenie, Moscow, 1985) p. 400
8. Noise meters: General technical requirements and test methods. State Standard GOST 17187-81 (Standard Publishing, Moscow, 1989), p. 28
9. P.N. Kravchun, *Generation and Methods of Abating Noise and Sound Vibration* (University Publishing, Moscow, 1991), p. 184
10. F.A. Jacobsen, V. Cutanda, P.M. Juhl, Sound intensity probe for measuring from 50 to 10 kHz. Bruel and Kjaer Tech. Rev. **1**, 1–8 (1996)
11. S. Gade, Sound intensity (part I theory). Bruel and Kjaer Techn. Rev. **3**, 3–39 (1982)
12. S. Gade, Sound intensity (part 2 instrumentation and applications). Bruel and Kjaer Tech. Rev. **4**, 3–32 (1982)
13. J.Y. Chung, Cross-spectral method of measuring acoustic intensity without error caused by instrument phase mismatch. J. Acoust. Soc. Am. **64**(6), 1613–1616 (1978)
14. F.J. Fahy, Measurement of acoustic intensity using the cross-spectral density of two microphone signals. J. Acoust. Soc. Am. **62**(4), 1057–1059 (1977)
15. G.C. Steyer, R. Singh, D.R. Houser, Alternative spectral formulation for acoustic velocity measurement. J. Acoust. Soc. Am. **81**(6), 1955–1961 (1987)
16. J.J. Christensen, J. Hald, Beamforming. Bruel and Kjaer Tech. Rev. **1**, 1–50 (2004)
17. J. Hald, Combined NAH and beamforming using the same arra. Bruel and Kjaer Techn. Rev. **3**, 3–39 (2005)
18. J. Patch Hald, Nearfield acoustical holography using a new statistically optimal method (SONAH). Bruel and Kjaer Tech. Rev. **3**, 40–52 (2005)
19. Brake squeal investigations using acoustic holography. case study. Brüel and Kjær sound and vibration measurement A/S, [Electronic resource]. http://www.bksv.com/pdf/ba0618.pdf, Accessed 4 Mar 2010

# Chapter 4
# Methods of Analysis of Noise and Vibration Signals

This chapter presents the general methods of the frequency analysis of vibration and noise signals, namely: Fourier transform, the analysis of modulated signals, the spectral analysis of random processes, the analysis of the interrelation (coherence) between two processes, and cepstral analysis. The methods of realizing the frequency methods in digital systems are considered too.

## 4.1 Methods of Frequency Analysis

### *4.1.1 Expansion in Fourier Series*

Expansion of a complex vibration process into the simplest components is called a frequency analysis. The value that indicates the behavior of the noise or vibration energy distribution within the frequency range is termed a frequency spectrum.

All vibration processes can be subdivided into the periodic and aperiodic ones. A periodic signal of the type $x(t) = x(t + T)$, where $T$ is a repetition period, is most characteristic for noise and vibration processes. It is very important to understand the behavior of the harmonic signal $x(t) = A\cos(\omega t - \phi)$, which is a function of three independent components, i.e., amplitude $A$, angular frequency $\omega$ and phase $\varphi$.

A harmonic signal can be presented as a trigonometric relation

$$A\cos(\omega t - \phi) = a\cos\omega t + b\sin\omega t, \tag{4.1}$$

where $A^2 = a^2 + b^2$; $\phi = arctg(b/a)$.



V.P. Sergienko and S.N. Bukharov, *Noise and Vibration in Friction Systems*,
Springer Series in Materials Science 212, DOI 10.1007/978-3-319-11334-0_4

Any periodic signal can be considered as a sum of harmonic components (harmonics), namely, a Fourier series

$$x(t) = \frac{a_0}{2} + \sum_{n=1}^{\infty} (a_n \cos n\omega_1 t + b_n \sin n\omega_1 t) = \frac{a_0}{2} + \sum_{n=1}^{\infty} A_n \cos(n\omega_1 t - \phi_n) \tag{4.2}$$

where the component $\frac{a_0}{2}$ defines the mean signal $x(t)$;

$A_n = \sqrt{a_n^2 + b_n^2}$ is the $n$-th harmonic amplitude;
$\phi = arctg(b_n/a_n)$ is the $n$-th harmonic phase;
$\omega_1 = 2\pi/T$—fundamental frequency of the signal.

The coefficients of Fourier series are related to the time function of a signal as follows $x(t)$:

$$\left.\begin{aligned} a_n &= \frac{2}{T}\int_{-T/2}^{T/2} x(t)\cos n\omega_1 t dt, n = 0, 1, 2, \ldots \\ b_n &= \frac{2}{T}\int_{-T/2}^{T/2} x(t)\sin n\omega_1 t dt, n = 0, 1, 2, \ldots \end{aligned}\right\}. \tag{4.3}$$

A set of amplitudes $A_n$ with multiple *to* each other frequencies $n\omega_1$, is forming a certain evenly distributed amplitude spectrum with periodic vibrations $x(t)$, and a set of phases $\varphi_n$ forms a *phase spectrum.*

A number of applied problems on the analysis of noise and vibration use a complex form of signal recording. Let us present a harmonic signal in a complex form

$$x(t) = A\cos(\omega_1 t - \phi) = \mathrm{Re}\left[Ae^{j(\omega_1 t - \phi)}\right], \tag{4.4}$$

and the periodic signal will be presented by a Fourier series

$$x(t) = \sum_{-\infty}^{+\infty} C_n e^{jn\omega_1 t}, \tag{4.5}$$

where

$$C_n = \frac{1}{T}\int_0^T x(t)e^{-jn\omega_1 t} dt, \tag{4.6}$$

the complex Fourier series coefficients except for the $C_0$, are: $C_0 = a_0$; $|C_n| = \frac{\sqrt{a_n^2+b_n^2}}{2}$; $\phi_n = arctg(b_n/a_n)$.

### 4.1.2 The Integral Fourier Transform

The spectrum of aperiodic signals is continuous and contains all frequencies. So, Fourier transformation into a series is inapplicable in this case the. Therefore, to consider named processes in a frequency area one should present the functions for aperiodic processes via the Fourier integral. An aperiodic signal may be taken as a periodic one in the case of a limiting transfer from the Fourier series in supposition that the vibration period is increasing till infinity $T \rightarrow \infty$. It seems justifiable to present an aperiodic function in such a way since we may consider it as a limiting case of the periodic function with an infinite period.

So, if we substitute $C_n$ from (4.6) into (4.5) by letting period $T$, to infinity, we shall obtain that

$$x(t) = \frac{1}{2\pi} \int_{-\infty}^{+\infty} e^{j\omega t} d\omega \int_{-\infty}^{+\infty} x(t) e^{-j\omega t} dt \tag{4.7}$$

or

$$x(t) = \frac{1}{2\pi} \int_{-\infty}^{+\infty} S(j\omega) e^{j\omega t} d\omega, \tag{4.8}$$

where

$$S(j\omega) = \int_{-\infty}^{+\infty} x(t) e^{-j\omega t} dt. \tag{4.9}$$

Quantity $S(j\omega)$ is a complex distribution function of the amplitudes or a *spectral density* of a spectrum:

$$S(j\omega) = |S(j\omega)| e^{j\psi}. \tag{4.10}$$

Let us find modulus $|S(j\omega)|$ and argument $\psi$ by presenting (4.7) in the form

$$x(t) = \frac{1}{2\pi} \int_{-\infty}^{+\infty} \left[ \int_{-\infty}^{+\infty} x(t) e^{j\omega(t-\tau)} d\tau \right] d\omega. \tag{4.11}$$

Substitution of (4.11) in the dependence

$$e^{j\omega(t-\tau)} = \cos\omega(t-\tau) + j\sin\omega(t-\tau)$$

will give us

$$x(t) = \frac{1}{2\pi}\left\{\int_{-\infty}^{+\infty}\left[\int_{-\infty}^{+\infty} x(\tau)\cos\omega(t-\tau)d\tau\right]d\omega + \int_{-\infty}^{+\infty}\left[\int_{-\infty}^{+\infty} x(\tau)\sin\omega(t-\tau)d\tau\right]d\omega\right\}. \tag{4.12}$$

The second integral in (4.12) equals to zero, i.e.,

$$\int_{-\infty}^{+\infty} x(\tau)d\tau \int_{-\infty}^{+\infty} \sin\omega(t-\tau)d\omega = \int_{-\infty}^{+\infty} x(\tau)\cos\omega(t-\tau)\Big|_{-\infty}^{+\infty} d\tau = 0,$$

therefore,

$$x(t) = \frac{1}{\pi}\int_{\omega=0}^{\omega=\infty}\int_{\tau=-\infty}^{\tau=\infty} x(\tau)\cos\omega(t-\tau)d\tau d\omega \tag{4.13}$$

The right-hand side of (4.13) is a Fourier integral. Let us express the integrand relation of the Fourier integral (3.13) as

$$\begin{aligned}\int_{-\infty}^{+\infty} x(\tau)\cos\omega(t-\tau)d\tau &= \cos\omega t\int_{-\infty}^{\infty} x(\tau)\cos\omega\tau d\tau + \sin\omega t\int_{-\infty}^{\infty} x(\tau)\sin\omega\tau d\tau \\ &= a'\cos\omega t + b'\sin\omega t\end{aligned} \tag{4.14}$$

where

$$a' = \int_{-\infty}^{\infty} x(\tau)\cos\omega\tau d\tau$$

$$b' = \int_{-\infty}^{\infty} x(\tau)\sin\omega\tau d\tau.$$

So, (3.13) with account of (3.14) will look like

$$x(t) = \frac{1}{\pi}\int_{0}^{\infty}(a'\cos\omega t + b'\sin\omega t)d\omega \tag{4.15}$$

or

$$x(t) = \frac{1}{\pi}\int_{0}^{\infty}A\sin[\omega t + \phi(\omega)]d\omega, \tag{4.16}$$

where $A_n = \sqrt{a'^2 + b'^2}$—amplitude and $\phi = arctg(b'/a')$ is a spectral density phase of the signal.

If to compare the results of (4.8) and (4.16), we shall come to a conclusion that

$$S(\omega) = A;\ \psi = \phi(\omega) \tag{4.17}$$

This means that the spectrum of aperiodic signals is characterized rather by spectral density $S(j\omega)$ than by the amplitude as the periodic signals do.

Equations (4.8) and (4.9) are the basic formulas for the theory of spectra expressed through a pair of Fourier transforms (4.8) interconnecting functions $x(t)$ and $S(j\omega)$. Using the inverse Fourier transform (4.8) and remembering its spectral density $S(j\omega)$ it is possible to reconstruct the signal waveform.

The integral Fourier transform can be simplified if to take the time reference of the signal in the form of either even $x_e(t)$ or odd $x_o(t)$ time function. So, the Fourier transform will take the form of either the first or the second relation, correspondingly:

$$\begin{aligned} S(j\omega) &= 2\int_{0}^{\infty}x_e(t)\cos\omega t dt; \\ S(j\omega) &= 2\int_{0}^{\infty}x_o(t)\sin\omega t dt. \end{aligned} \tag{4.18}$$

The main spectral characteristics of the impulse signals are the spectral energy [1]

$$W(\omega) = \int_{0}^{\infty}|S(j\omega)|^2 d\omega. \tag{4.19}$$

and the energy spectrum

$$N(\omega) = \frac{1}{\pi} \lim_{T \to \infty} \frac{|S(j\omega)|^2}{T}. \tag{4.20}$$

The active spectrum width is determined as a frequency band, and the mcorresponding part λ of the total impulse energy is found. The value of Δω is obtained from the relation:

$$\lambda \int_0^{\infty} |S(j\omega)|^2 d\omega = \int_0^{\Delta\omega} |S(j\omega)|^2 d\omega. \tag{4.21}$$

Spectral representation of impulse processes gives a perfect result if its is important to simulate all components of the impulse frequency. A spectrum (spectral density versus frequency) of a shock pulse can be given in the form of either a displacement, velocity or acceleration spectrum. The latter two spectra are obtained by multiplying the displacement spectrum by ω and $\omega^2$.

Figure 4.1a shows an example of a bell-shaped pulse spectrum $x(t) = Ae^{-\beta t^2}$ with a form factor β. Its signal spectrum is described by the following relation

$$S(j\omega) = 2A \int_0^{\infty} e^{-\beta t^2} \cos \omega t dt = A\sqrt{\frac{\pi}{\beta}} \cdot e^{-\frac{\omega^2}{4\beta}}. \tag{4.22}$$

The spectral analysis of such signals has shown that the less the signal pulse duration, i.e., the larger is the form factor β, the broader is the spectrum and the less is the spectral density of the signal at low frequencies. In the HF range the spectral

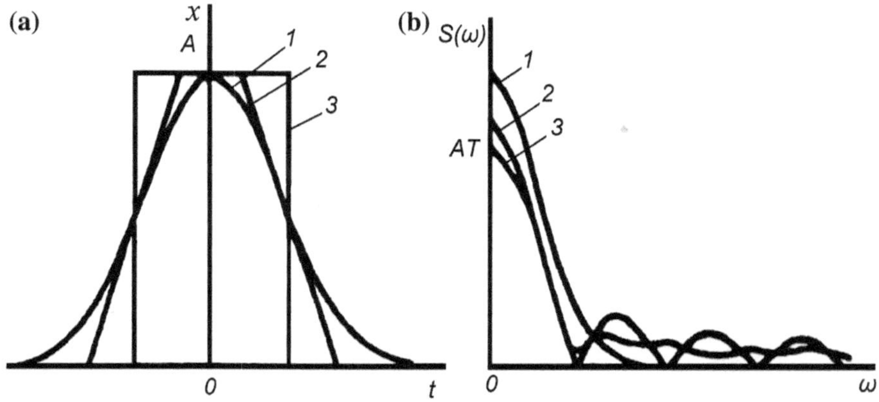


**Fig. 4.1** Form (**a**) and spectra (**b**) of the pulses: *1*—bell-shaped, *2*—trapezoidal, *3*—square

density value of a signal is strongly affected much not only by the pulse duration but also by steepness of the pulse edge. Figure 4.1b illustrates a trapezoidal and square pulses, which spectra are found by the corresponding relations

$$\begin{aligned} S_{mp}(j\omega) &= \left[\frac{2A}{\omega(T - T_1)}\right](\cos\omega T_1 - \cos\omega T), \\ S_{np}(j\omega) &= \frac{2A}{\omega}\sin\frac{\omega T}{2}, \end{aligned} \tag{4.23}$$

where $A$ is the maximal pulse amplitude; $T$ is pulse duration, and $T_1$—time of trapezoidal pulse vertex.

The comparison of the bell-shaped, trapezoidal and square pulse spectra of a similar amplitude and effective time $T'_{eff} = \frac{1}{A}\int_0^T x(t)dt$ within the HF range has shown that the steep-edged pulses display a higher spectral density.

### 4.1.3 Analysis of Modulated Signals

There exists a possibility of simultaneous existence and interaction of complex vibration signals in mechanical systems.

In a simplest case, a combination of two vibrations with frequencies $\omega_0$ and $\Omega$, and two phases $\phi$ and $\Psi$ is of the form:

$$x(t) = A\cos(\omega_0 t + \phi)\cos(\Omega t + \Psi) \tag{4.24}$$

The analysis of such vibroacoustic waveforms requires a model enabling to divide a signal into simple components. Expansion into the Fourier series describes the signal as an additive mixture of two harmonic signals with the amplitude $A/2$ and frequencies $\omega_0 \pm \Omega$. In this case, the signal will be of the form:

$$x(t) = \frac{A}{2}\{\cos[(\omega_0 + \Omega)t + \varphi + \psi] + \cos[(\omega_0 - \Omega)t + (\varphi - \psi)]\} \tag{4.25}$$

Modulation processes of vibroacoustic signals carry valuable information on the technical state of the mechanical system under study. In this connection, we should consider them in more detail.

Let us express a modulated signal in the form of an additive multiplicative mixture of signals $u(t)$ and $v(t)$:

$$x(t) = A[a_1 u(t) + a_2 v(t) + a_3 u(t)v(t) + \cdots] \tag{4.26}$$

where $a_1$, $a_2$, $a_3$, … are the constants proportional to signal amplitudes.

A simplest amplitude-modulated signal is determined for the case of harmonic signals $u(t) = \cos\omega_0 t$ and $v(t) = \cos\Omega t$ based on the relation [2, 3]:

$$\begin{aligned} x(t) &= A[1 + m\cos(\Omega t + \phi)]\cos\omega_0 t \\ &= A\cos\omega_0 t + \frac{1}{2}A\cos[(\omega_0 - \Omega)t - \phi] + \frac{1}{2}Am\cos[(\omega_0 + \Omega)t + \phi] \end{aligned} \tag{4.27}$$

where $m = \frac{x_{\max} - x_{\min}}{x_{\max} + x_{\min}}$—modulation depth.

Proceeding from Fig. 4.2, the signal contains three harmonic components, namely, a carrier with $\omega_0$ frequency and two sidetones with frequencies, $\omega_0 + \Omega$ and $\omega_0 - \Omega$. A beat-frequency waveform containing two harmonic components described by (4.26) is shown for comparison in Fig. 4.2b. Function $A[1 + m\cos(\Omega t + \phi)]\cos\omega_0 t$ describes a modulated waveform *envelope*. The difference between the side tone phases amounts to a starting phase that tends to $\phi \ll \frac{\pi}{2}$ at $\Omega \ll \omega_0$, which is further omitted due to its smallness.

The amplitude modulation with a random modulating signal $\phi(t)$ is characterized by the relation

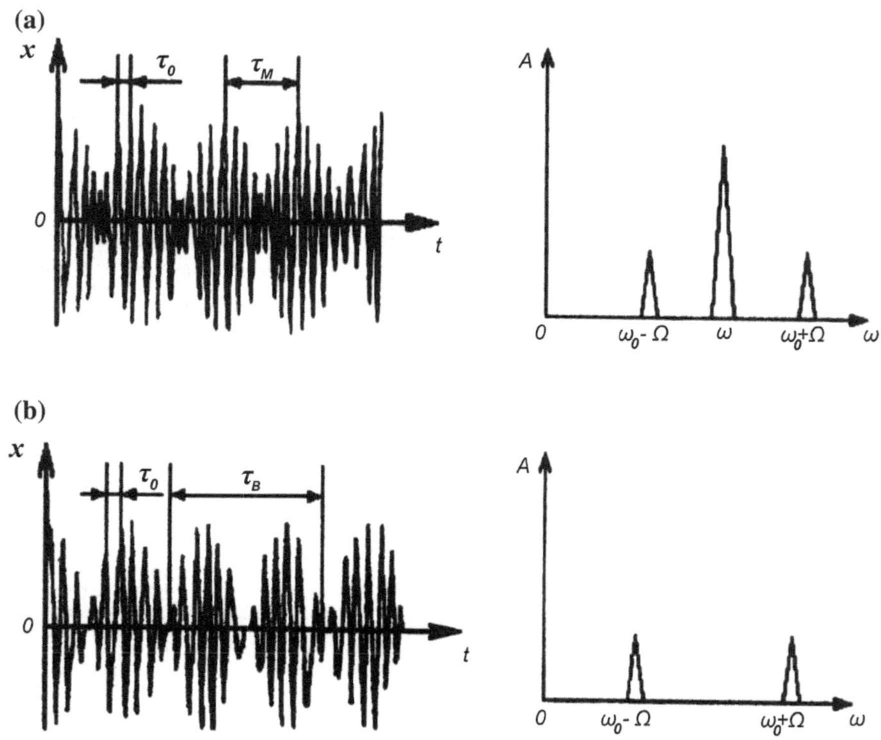


**Fig. 4.2** Form and spectrum of amplitude-modulated signal (**a**) and beat-frequency waveform (**b**): $\tau_0 = 2\pi/\omega_0$—carrier period; $\tau_M = 2\pi/\Omega$—modulation period; $\tau_B = \pi/\Omega$—beat period

$$x(t) = A\left[1 + m\sum_{k=1}^{n} a_k \cos(\Omega_k t - \phi_k)\right] \cos \omega_0 t \tag{4.28}$$

where $ma_k = m_k$—partial modulation depth.

To understand the frequency modulation in terms of physical notions, let us express a harmonic signal as follows

$$x(t) = \cos(\omega_0 t + \phi_0) = A \cos[\phi(t)] \tag{4.29}$$

where $\phi(t) = \omega_0 t + \phi_0$—total phase of harmonic signal.

In a general case, signal frequency $\omega(t)$ may vary randomly. In the case of harmonic variations of the frequency, the total phase is of the type [2–4]:

$$\phi(t) = \omega_0 t + m\frac{\omega_0}{\omega_m} \sin \omega_m t + \phi_0 \tag{4.30}$$

In this case, the signal (4.29) is a frequency-modulated one with modulation depth $m$, frequency deviation $m\omega_0$ and modulation index $\beta = \frac{m\omega_0}{\omega_m}$.

We may, possibly, accept (4.30) as a phase modulation with depth $\frac{m\omega_0}{\omega_m}$, although the original relation (4.30) is defining the frequency modulation. Indeed, $\omega = \frac{d\phi}{dt}$ and division into the phase and frequency modulation of the modulated signal is only conventional. In the discourse to follow these two modulation types are joined into one term “angular modulation”.

A signal with an angular harmonic modulation of the type

$$x(t) = A \cos(\omega_0 t + \beta \sin \omega_m t + \phi_0) \tag{4.31}$$

for $\phi_0 = 0$ and small modulation indices $\beta \ll 1$, under which $\sin(\beta \sin \omega_m t) \cong \beta \sin \omega_m t$, is characterized by a spectrum

$$\begin{aligned} x(t) &= A(\cos \omega_0 t - \beta \sin \omega_m t \sin \omega_0 t) \\ &= A \cos \omega_0 t + \frac{\beta}{2} A \cos(\omega_0 + \omega_m)t - \frac{\beta}{2} A \cos(\omega_0 - \omega_m)t. \end{aligned} \tag{4.32}$$

If to compare the amplitude-modulated signal spectrum (4.27) to the angular modulated one (4.32), we shall see that the difference is in the side tone phases of the frequencies $\omega_0 \pm \omega_m$. The sidetone phase differs by $\phi \leq \frac{\pi}{2}$, for the amplitude-modulated signals and by $\pi$ for the angular-modulated signals. A signal of angular modulation with large modulation indices looks like:

$$x(t) = A\left\{J_0(\beta) \cos \omega_0 t + \sum_{n=1}^{\infty} J_n(\beta)[\cos(\omega_0 + n\omega_m)t + (-1)^n \cos(\omega_0 - n\omega_m)]\right\} \tag{4.33}$$

where $J_n(\beta)$ is the $n$-th order Bessel function of the first kind.

It follows from (4.33) that the angular-modulated signal spectrum is wider than the amplitude-modulated signal ones. Notice that the efficient bandwidth depends on the modulation index and is roughly equal to $2\beta\omega_m$ [15].

### 4.1.4 Spectral Analysis of Random Signals

Such spectral characteristics as a spectral power density $G(\omega)$, spectral function $F_s(\omega)$, bandwidth $\Delta\omega$, maxima location and magnitude of a spectral power density $G(\omega)$, boundary frequencies $\omega_1$ and $\omega_2$, that correspond to a given level of $G(\omega)$ reduction relative to a maximal value of $G(\omega)$ are widely used along with the previously described ones (see Chap. 1) used to analyze random processes.

Spectral density can be found from Wiener-Hinchin's relations in the form of Fourier transform of a corresponding correlation function, $R(\tau)$, i.e.,

$$\begin{aligned} G(\omega) &= \frac{1}{2\pi} \int_{-\infty}^{\infty} R_x(\tau) e^{-j\omega\tau} d\tau, \\ R_x(\tau) &= \int_{-\infty}^{\infty} G(\omega) e^{j\omega\tau} d\omega. \end{aligned} \tag{4.34}$$

Function $G(\omega)$ denotes a random process power per frequency band $d\omega$. In this connection, $G(\omega)$ is also called the energy spectrum of the signal being studied.

The process power enclosed in the frequency band between $\omega_1$ and $\omega_2$ is defined by the relation

$$P_{12} = \int_{\omega_1}^{\omega_2} G(\omega) d\omega. \tag{4.35}$$

To understand spectral function $F(\omega)$ let us express the correlation function $R_x(\tau)$ in the form

$$R_x(\tau) = \int_{-\infty}^{\infty} e^{j\omega\tau} dF(\omega), \tag{4.36}$$

where $F(\omega)$ is a real nondecreasing bounded function of argument $\omega$.

Function $F_s(\omega)$ is a spectral function. In the case

$$\int_{-\infty}^{\infty} |R_x(\tau)|d\tau < \infty,\ G(\omega) = \frac{dF(\omega)}{d\omega},\ \text{then} \qquad (4.37)$$
$$\Delta F(\omega_i) \approx G(\omega_i)\Delta\omega_i.$$

Proceeding from the parity property of functions $R_x(\tau)$ and $G(\omega)$, (4.34) can be written as follows

$$G(\omega) = \frac{1}{\pi}\int_{0}^{\infty} R_x(\tau)\cos\omega\tau d\tau,$$
$$R_x(\tau) = 2\int_{-\infty}^{\infty} G(\omega)\cos\omega\tau d\omega. \qquad (4.38)$$

One should remember that if $\tau = 0$, then

$$R_x(0) = P_\infty = \int_{0}^{\infty} G(\omega)d\omega, \qquad (4.39)$$

where $R_x(0)$ is the variance of a random process $x(t)$. Relation (4.39) expresses a total power $P_\infty$ of the process $x(t)$.

Spectral density can be expressed through a current spectrum of realizations [5]:

$$S_t(\omega) = \int_{0}^{t} x(t)e^{-j\omega t}dt.$$

To this end, let us set-up an equation for the energy of the process $x(t)$ generated during time $t$,

$$E_t = \int_{0}^{t} x^2(t)dt = \frac{1}{\pi}\int_{0}^{\infty} |S_t(\omega)|^2 d\omega. \qquad (4.40)$$

The mean power per time $t$ is

$$P_t = \frac{E}{t} = \frac{1}{\pi t}\int_{0}^{\infty} |S_t(\omega)|^2 d\omega. \qquad (4.41)$$

For a stationary process $x(t)$ the mean power is found from relation

$$P_{\infty} = \lim_{t\to\infty} P_t = \frac{1}{\pi} \lim_{t\to\infty} \frac{1}{t} \int_0^{\infty} |S_t(\omega)|^2 d\omega. \tag{4.42}$$

It becomes evident if we compare (4.39) and (4.42) that

$$G(\omega) = \frac{1}{\pi} \lim_{t\to\infty} \frac{|S_t(\omega)|^2}{t}. \tag{4.43}$$

There is one more form of the link between spectral density $G(\omega)$ and the current spectrum of realization $S_t(\omega)$

$$G(\omega) = \frac{1}{\pi} M\left(\frac{\partial}{\partial t} |S_t(\omega)|^2\right), \tag{4.44}$$

where $M$—mathematical expectation with respect to a set.

To make a spectral analysis of the signals of random processes one should take into account the relation of spectrum width $\Delta f$ to the correlation interval $\Delta\tau$

$$\Delta f \Delta\tau = 1. \tag{4.45}$$

Quantities $\Delta\omega$ and $\Delta\tau$ express the efficient length of the functions of spectral density $G(\omega)$ and $R_x(\tau)$ correlation, correspondingly.

Equation (4.45) can be explained as follows. In relations (4.38) we accept $\omega = 0$ in the first equation, and $\tau = 0$ in the second one. Consequently,

$$G(0) = \frac{1}{2\pi} \int_{-\infty}^{\infty} R_x(\tau) d\tau,$$

$$R_x(0) = \int_{-\infty}^{\infty} G(\omega) d\omega. \tag{4.46}$$

The integrals in (4.46) present in fact the areas under curves $G(\omega)$ and $R_x(\tau)$ respectively. With account of the length of functions $G(\omega)$ and $R_x(\tau)$, the quantities $\Delta\omega$ and $\Delta\tau$ will be

$$\Delta\omega = \frac{1}{G(0)} \int_{-\infty}^{\infty} G(\omega) d\omega$$

$$\Delta\tau = \frac{1}{R_x(0)} \int_{-\infty}^{\infty} R_x(\tau) d\tau \tag{4.47}$$

By considering relations (4.46) we obtain (4.45) from (4.47).

Consequently, relations (4.47) can be understood so that the areas under curves $G(\omega)$ and $R_x(\tau)$ are equal to those of the rectangles with the bases $\Delta\omega$ and $\Delta\tau$, and heights $G(0)$ and $R_x(0)$, correspondingly.

When analyzing the stationary random processes it is important to present them from the viewpoint of the frequency range, i.e., in the form of the narrow-band and broad-band ones.

It is typical for the narrow-band vibration processes to concentrate the main portion of its energy in one or a few relatively narrow frequency bands. The amplitudes of such processes are varying randomly. The narrow-band vibrations can be presented analytically in the form:

$$x(t) = x_a(t)\sin[\omega_i t + \phi(t)]. \tag{4.48}$$

where $x_a(t)$ and $\phi(t)$ are slowly varying functions in contrast to $\sin\omega_i t$.

The narrow-band vibrations are characteristic for the vibration systems with one degree of freedom that experience the effect of a broad-band random vibration. The narrow-band vibrations look like the harmonic ones and are therefore called the *almost-harmonic* vibrations.

The correlative functions and spectral densities of the narrow-band vibration are of the form

$$\begin{aligned} R_{1x}(\tau) &= e^{-\alpha|\tau|}\cos\omega_0\tau, \\ R_{2x}(\tau) &= e^{-\alpha|\tau|}(\beta\cos\omega_0\tau + \gamma\sin\omega_0|\tau|), \end{aligned} \tag{4.49}$$

$$\begin{aligned} G_1(\omega) &= 2\alpha\left[\frac{1}{\alpha^2 + (\omega+\omega_0)^2} + \frac{1}{\alpha^2 + (\omega-\omega_0)^2}\right], \\ G_2(\omega) &= 2\alpha\left[\frac{2\alpha\beta + 2\gamma(\omega+\omega_0)}{\alpha^2 + (\omega+\omega_0)^2} + \frac{2\alpha\beta - 2\gamma(\omega-\omega_0)}{\alpha^2 + (\omega-\omega_0)^2}\right], \end{aligned} \tag{4.50}$$

where $\alpha$, $\beta$, $\gamma$—constants.

The broad-band vibration is composed of a sum of several narrow-band oscillation processes and vibration-excited noise $n(t)$

$$x(t) = \sum_{i=1}^{n} x_{ai}(t)\sin[\omega_{0i}t + \phi_i(t)] + n(t). \tag{4.51}$$

Vibrational noise $n(t)$ presents oscillations with a mixed modulation (both amplitude and angular) [6]. The vibration noise includes a great number of low-intensity constituents.

The broad-band vibroacoustic phenomena in friction joints are presenting mostly the random broad-band processes. Their frequency range is varying within 0–20 kHz. Since the broad-band processes include a combination of the narrow-band ones, their correlative function and spectral density are equal to a sum of the corresponding functions of the form (4.49) and (4.50). Named functions describe noise and vibration of numerous objects quite perfectly.

### 4.1.5 Cross-Correlation Function of Two Processes

The spectral properties of a single-time function $x(t)$ are characterized by the spectral density of energy $G_{xx}(f)$, also called as an *auto-spectral density* or a *power spectrum*. Correspondingly, the power spectrum of the two time functions $x(t)$ and $y(t)$ is called a cross-spectral density $G_{xy}(f)$ or $G_{yx}(f)$, also termed a cross-spectrum.

$$G_{xx}(f) = F_x(f)F_x^*(f) = |F_x(f)|^2; \tag{4.52}$$

$$G_{xy}(f) = F_x(f)F_y^*(f); \tag{4.53}$$

$$G_{yx}(f) = F_y(f)F_x^*(f), \tag{4.54}$$

where $F_x(f)$ and $F_y(f)$ are the direct Fourier transform for functions $x(t)$ and $y(t)$, $F_x(f)$ and $F_y(f)$ present a complex-conjugate function.

The functions of spectral density are related with the functions in the time domain by the following equations

$$G_{xx}(f) = 2\int_{-\infty}^{+\infty} R_{xx}(\tau)e^{-j2\pi f\tau}d\tau, \quad 0 \le f \le +\infty; \tag{4.55}$$

$$R_{xx}(f) = \frac{1}{2}\int_{-\infty}^{+\infty} G_{xx}(\tau)e^{-j2\pi f\tau}df, \quad -\infty \le \tau \le +\infty; \tag{4.56}$$

$$G_{xy}(f) = 2\int_{-\infty}^{+\infty} R_{xy}(\tau)e^{-j2\pi f\tau}d\tau, \quad 0 \le f \le +\infty; \tag{4.57}$$

$$R_{xy}(f) = \frac{1}{2}\int_{-\infty}^{+\infty} G_{xy}(\tau)e^{-j2\pi f\tau}df, \quad -\infty \le \tau \le +\infty, \tag{4.58}$$

where $R_{xx}(\tau)$ and $R_{xy}(\tau)$ are the autocorrelation and cross-correlative functions.

The cross-correlation coefficient $\rho_{xy}(\tau)$ is of the kind

$$\rho_{xy}(\tau) = \frac{R_{xy}(\tau)}{\sqrt{R_{xx}(0)R_{yy}(0)}}. \tag{4.59}$$

The transfer function is found by solving next relations

$$H(f) = \frac{S_y(f)}{S_x(f)};\ |H(f)|^2 = \frac{G_{yy}(f)}{G_{xx}(f)};\ H(f) = \frac{G_{yx}(f)}{G_{xx}(f)}. \tag{4.60}$$

It should be emphasized that most complete information on the phase and amplitude is obtained by using a cross-spectral density.

The *coherent function* $\gamma_{xy}^2(f)$ or so-called normalized cross-spectrum is analogous to the correlation coefficient within the frequency range. It reflects the degree of interrelation linearity between the harmonic components of vibration processes:

$$\gamma_{xy}^2(f) = \frac{|G_{xy}(f)|^2}{G_{xx}(f)G_{yy}(f)}. \tag{4.61}$$

The coherent function for a linear system is identically equal to a unit that diminishes with increasing nonlinearity degree of the system $\gamma_{xy}^2(f)$, which means that $0 \le \gamma_{xy}^2(f) \le 1$. The coherent function is used as a measure of nonlinearity, especially when the correlation coefficient turns to be non-informative for being defined in the frequency band, whereas the coherent function is determined at each frequency.

### 4.1.6 Cepstral Analysis

The real vibration signals (noise) are measured in a certain point of an object (space). They consist as a rule of a mixture of multiple-reflected from the surface signals and the ones that have undergone changes due to the resonant properties of the object elements being studied.

These signals constitute a convolution of signals created by their source and an impulse response of the mechanical system between the points of vibration excitement and measurement.

The equation for the convolution of signals looks like

$$x(t) = \int_{-\infty}^{+\infty} x_0(\tau)g(t-\tau)d\tau, \tag{4.62}$$

where $x(t)$ is a signal in the measurement point; $x_0(\tau)$ is a signal in the point of excitement; $g(t-\tau)$ is the impulse response of the vibration system.

The cepstral transformation is used to separate such complex signals and isolate the constituent created by the vibration source itself. A cepstrum is a term formed by exchanging letters in the word spectrum, which means a Fourier transform square of the logarithmic spectrum of a signal power:

$$K(\tau) = \{F[\lg G(\omega)]\}^2 = \left\{ \int_0^{\infty} \lg [G(\omega)]^2 \cos \omega\tau d\omega \right\}^2. \tag{4.63}$$

Figure 4.3 shows the direct frequency spectra of a vibration signal and its cepstrums [7].

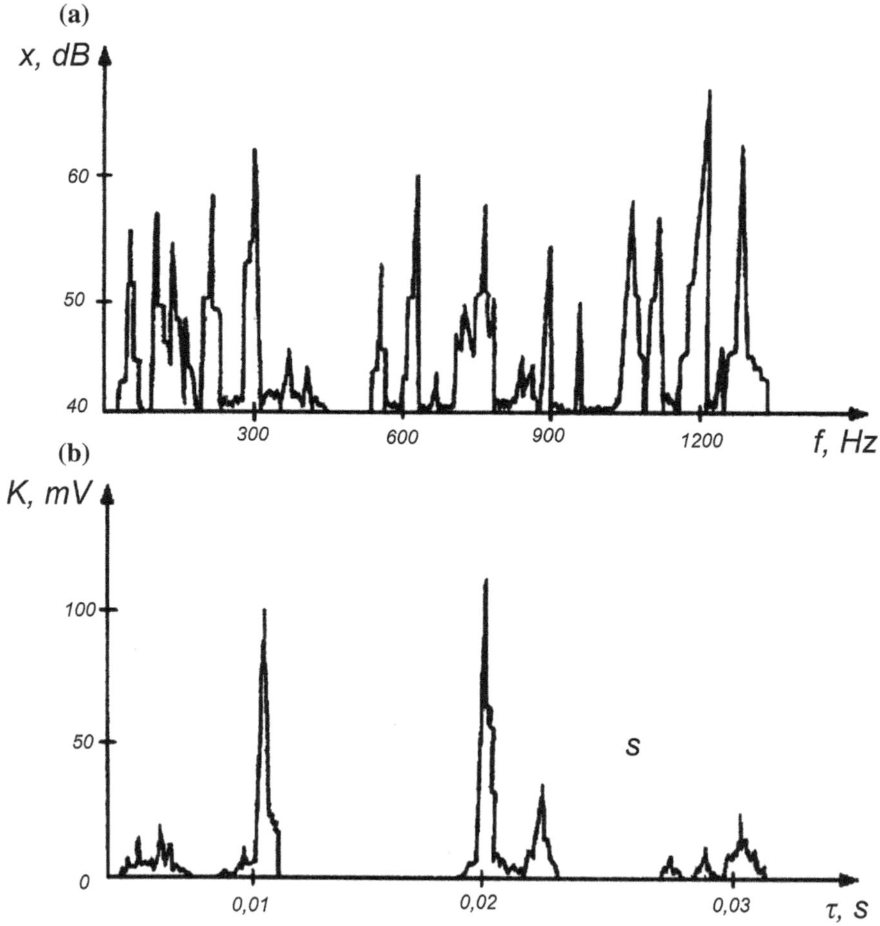


**Fig. 4.3** A spectrum (**a**) and a cepstrum (**b**) of vibration signals of a mechanism with rotating elements

The energy of a vibroacoustic signal scattered over numerous harmonics is localized in its spectral representation only in one component in the case of a cepstral representation of the signal. A cepstrum is an efficient tool for analyzing signals with the amplitude and angle modulations. In particular, a cepstrum assists in determining modulation depth of just as random so complex in form signals. The traditional analysis of such signals requires a tedious and complex computation of the partial modulation depth.

Above-presented brief analysis of the models of complex signals proves that every type of the noise or vibration signals should be treated by a specific method in order to separate them into simple components. The Fourier transform is widely applied to expand the studied signals over the orthogonal basis of harmonic functions due to its convenience in technical realization by the analogue facilities. Along with above-named methods, modern digital procedures of signal processing offer some other orthogonal functions to decompose a complex signal into the simpler ones. The choice of the basis depends on available means of signal representation. This is why, particular problems of signal analysis, e.g., in vibrodiagnostics, the researchers use the orthogonal functions as the basis since they furnish information on vibration sources in a simplest and most convenient form.

## 4.2 Frequency Analysis Realized by Digital Devices

Rapid pace of the digital computing techniques, availability of the high-speed input/output devices of analogue information, along with a low-speed vibration spreading and restricted frequency range of processed signals contribute into the efficient computer analysis of vibration signals. The possibility of flexible changes in signal-processing algorithms and accelerated analysis have lead to a situation that despite their intricacy, the digital analyzing devices tend to drive out the analogue means. What is more, the algorithms can be presented in combination as a software package that enables to refuse from the analyzers in scientific research.

A signal analyzer includes a computer and peripheral input-output devices that are transforming the analogue signals into a numeric form, transfer them to a PC and store there. The division of a digital analyzer into the blocks shown in Fig. 4.4 is made conventionally to visualize the chief stages of the analogue signal processing. The peripheral facilities, such as input band-pass filter 1 usually operate

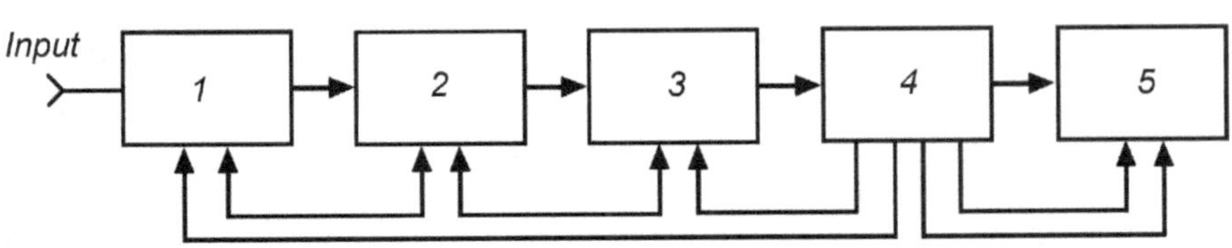


**Fig. 4.4** Diagram of a PC-aided digital signal analyzer

from the control signals of PC 4. However, in a number of cases the analogue-to-digital conversion 2 and accumulation of the transformed signal in the storage device 3 can be autonomous. This gives additional advantages in the analogue signal processing by the PC (5—input device).

A continuous analogue signal is transformed into a digital form by the analogue-to-digital converter (ADC) that makes time digitizing and amplitude quantification of the signal. The basis of the one-to-one transformation is a right choice of digitizing frequency $f_d = \omega_d/2\pi$ and amplitude quantification levels $N_{max}$.

According to Kotelnikov theorem, the correct one-to-one transformation of continuous signals into a digital form is ensured when

$$f_d \geq \Delta f_e, \tag{4.64}$$

where $\Delta f_e = \Delta\omega/2\pi$ is the efficient bandwidth of a signal spectrum. It should be noted that in a general case, the lower limit of the signal spectrum can have a frequency different from zero.

Violation of conditions (4.64) misalignes the analogue and digital forms of presenting a signal. This is why, one and the same chosen signal values may correspond to a group of analogue signals with a different harmonic composition. So-called overlapping effect of the frequencies is illustrated in Fig. 4.5 [7]. Figure 4.5a shows two harmonic components of the studied signal with an efficient band $\Delta f_e$, having similar amplitudes and frequencies corresponding the upper and lower limits of the signal $f_l$ and $f_u$. When a signal discretization frequency is $f_d \leq 2(f_u - f_l)$, then the HF component of the signal can be lost upon restoration of the signal (Fig. 4.5b).

It should be underlined that the analogue and digital components of signal representation may be not always aligned, even if condition (4.64) is observed. Figure 4.5c shows the harmonic components of the signal with frequencies $f_u/f_l = 3$, and its discretization results at frequency $f_d = 2(f_u - f_l)$. The restored signal (Fig. 4.5d) is also devoid of the component with frequency $f_u$. To achieve unambiguous correspondence an additional condition should be met:

$$\frac{kf_d}{2} \leq f_l; \quad f_u \leq \frac{(k+1)}{f_d/2}, \tag{4.65}$$

where $k = 1, 2, \ldots$

According to (4.65) for the signals shown in Fig. 4.6c, we have that $f_d = 2(f_u - f_l) = 4f_l$. We cannot, however, use this frequency for discretization, since$f_d/2 = 2f_l > f_l$ and $f_d/2 < f_u$. For instance, when the signal has a spectrum width 2 kHz and the lower frequency limit $f_l$ = 0; 2; 4; 6 kHz we obtain that the minimal one is $f_d$ = 4 kHz; if $f_l$ = 1 kHz, then $f_d$ = 6 kHz; when $f_l$ = 3 kHz, so $f_d$ = 5.0 kHz, when $f_l$ = 5.0 kHz, then $f_d$ = 4.7 kHz.

The possibility to reduce frequency of discretization and, consequently, memory span of the storage unit (SU), and computation volume via narrowing the signal frequency band stored in the PC simplifies the requirements to the digital waveform

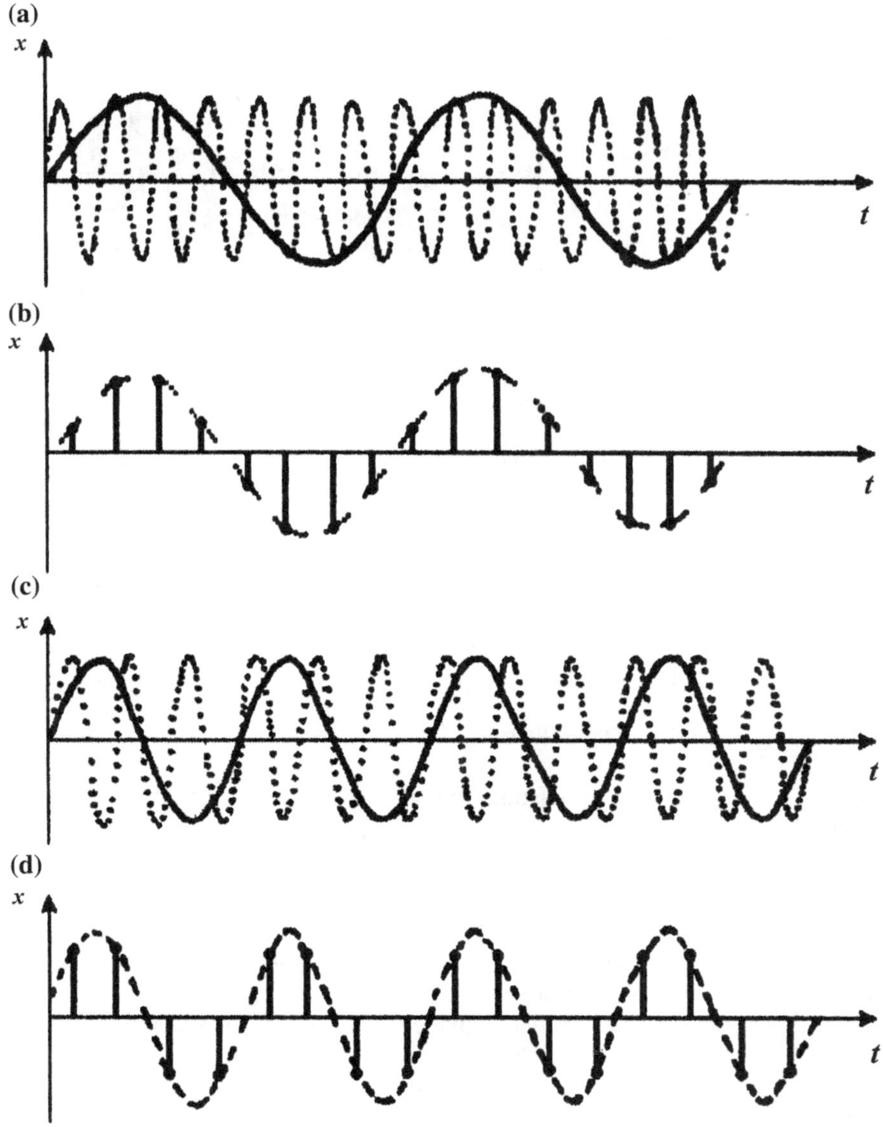


**Fig. 4.5** Harmonic waveforms: a discrete analogue (**a**, **c**) a restored signal from discrete one (**b**, **d**)

analyzers. One should supply the signal to the analyzer inlet through the band filters generating most informative for a certain problem part of a spectrum.

When a signal is level-quantified, its amplitude in the moment of discretization transforms into a digital form. The quantification range of a signal by its level is conditioned by the dynamic range of the signal in the vibration inverter or a microphone outlet fit with a preamplifier. The signals are most often input from a

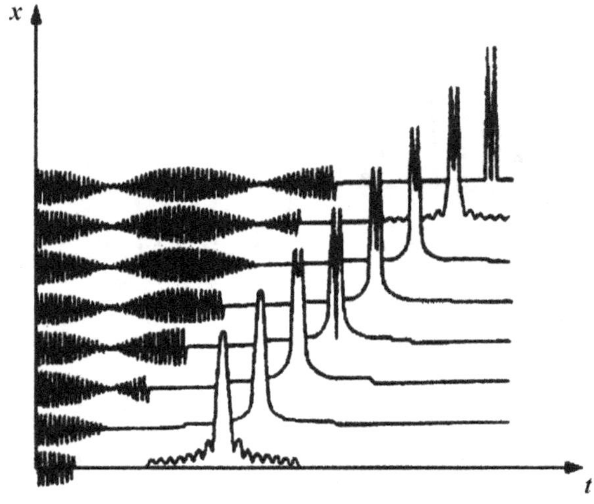


**Fig. 4.6** Form and spectrum of signal sections of different duration

PC by the devices with below 45–50 dB dynamic range in the up to 20 kHz band. So, to solve the problems of signal analysis we typically use the ADC with up to 60 dB dynamic range. It is to be noted that the technical specifications of ADC present also the data of the binary digits of the converter $m$ related to the dynamic range via the equation:

$$D_m = 20\lg N_{\max} = 20\lg(2^m - 1), \qquad (4.66)$$

where $N_{\max} = (2^m - 1)$ is a maximal number deduced by the ADC with binary digits.

A signal transformed in the ADC is sent to an SU by a PC instruction. The instructions for the ADC and other devices for information input/output are joined into a special program for communication with the PC, so-called drivers, available in a suite with mathematical support. However, special analytic problems require a separate development of such drivers.

The process of signal accumulation in the PC includes transformation of a signal amplitude into a code, determination of the word number in the SU memory (word address) and code recording in the indicated address. This process should proceed within the interval between the signal select.

The SU memory is characterized by a volume defined by discretization frequency $f_d$, dynamic range of the ADC, $D_m$ and the length of a signal section $\Delta T$ related to the resolving capacity of frequency $\Delta f_r$ by the equation

$$\Delta T = 1/\Delta f_r. \qquad (4.67)$$

The resolving capacity value is usually set according to the task of the vibration signal analysis, its aim and intricacy.

The memory span for the words, each corresponding to one signal which means transformed into a digital code, can be found from the relation:

$$N_c = \Delta T f_d, \tag{4.68}$$

and the physical volume of the memory $V_M$ equals to

$$V_M = N_c D_m = \Delta T f_d D_m. \tag{4.69}$$

The physical memory capacity is necessary for estimating the potentialities of the computer required to analyze a signal.

In modern computers with a bite structure of the SU (one bite equals to 8 binary digits) and dynamic range of the ADC about 6 dB, one word occupies two memory bites. Thus, to analyze a signal of 10 kHz spectrum width and 1 Hz filtration band (resolving capacity) the SU memory space for storing one signal section of length $\Delta T = 1$ s and discretization frequency $f_d$ = 20 kHz should be equal to 40 kB. To analyze a signal of 1 kHz band, the memory capacity needed to store a signal section reduces to 4 kB.

A number of computers uses a random-access storage unit (RSU). One RSU stores information from the ADC, another one sends information simultaneously to the main SU of the computer, for which aim a disc storage unit is used. This helps to eliminate discontinuities between sequential signal sections to be stored. High-speed computers with rather low discretization rate may produce a required noise and vibration analysis within the time less than the discretization interval $T_d = 1/f_d$. This type of analysis is called a real-time analysis, meaning that the output device, e.g., the observation device, reflects the results of signal analysis within any preceding storage time interval. Most of computers programmed in high-level languages show high enough speed of processing accumulated signals, so their analysis in real time is probable only for the LF signals. The speed of signal processing can be accelerated significantly by using programming languages of machine instructions [8].

Vast promises in solving problems of the waveform analysis in the real time show special microprocessors performing digital processing of a signal schematically. To solve a complex problem of analyzing vibration signals it is convenient to divide the data in time so that one data array can be processed simultaneously with accumulation of the next section of the signal. A computer SU can be conventionally subdivided in this case into three parts, namely: for accumulation of one signal section; for processing of previously stored signals; for processing program input of the vibration signal. So, the total memory space of a computer with separation in time for the spectral analysis of vibration within the frequency range 0–10 kHz and resolving capacity of frequency 1 Hz is about 120 kB. When analyzing vibration signals within 0–1 kHz frequency range, the memory space is reduced to 30 kB, and the processing program occupies about 20 kB of the memory space.

The main task of the computer-aided spectral analysis of vibration signals is a digital filterng based on the discrete Fourier transformation (DFT) for a parallel analysis of signal spectra.

In a general form, the DFT is a discrete analogue of a continuous transformation of the complex signals (4.5). It can be written with account of infinity of the signal frequency band $x(t)$ as follows:

$$x(t_i) \equiv x_i = \sum_{n=-(N-1)/2}^{(N-1)/2} C_n e^{jn\omega_1 t_i} = \sum_{n=-(N-1)/2}^{(N-1)/2} C_n e^{jn\frac{2\pi}{N}i}, \tag{4.70}$$

$$C_n = \frac{1}{N}\sum_{i=0}^{N-1} x(t_i)e^{-in\omega_1 t_i} = \frac{1}{N}\sum_{i=0}^{N-1} x_i e^{-jn\frac{2\pi}{N}} \tag{4.71}$$

where $C_n$—complex coefficients of the Fourier series; $N$—number of equidistant time moments of discretization within a signal period $x(t)$, $\omega_1 t_i = 2\pi i/N$—total signal phase.

The DFT with account of periodic properties of the complex coefficients will look like

$$x_i = \sum_{n=0}^{N-1} C_n e^{jn\frac{2\pi}{N}i}, \tag{4.72}$$

$$C_n = \frac{1}{N}\sum_{i=0}^{N-1} x_i e^{-jn\frac{2\pi}{N}i}. \tag{4.73}$$

The complex coefficients of the DFT can be presented in the form:

$$C_n = (a_n + jb_n)/2. \tag{4.74}$$

For a material signal $x(t)$ the DFT coefficients $C_n$ display the properties of Hermitian conjugation: $C_n^{\ni} = (a_n - jb_n)/2 = C_{-n}$. Therefore, coefficients $a_n$ and $b_n$ are symmetric relative to $n = N/2$, therefore to acquire full information on the transformed section of signal $x(t)$ it is suffice to input half of coefficients $a_n$ and $b_n$ into the SU. Such a property enables to use so-called compact algorithm of the DFT.

It is generally accepted to present DFT in a matrix form when analyzing signals by computer. A combination of values $\{x_i\}$ corresponds to an $N$-dimensional vector $\overline{X}$, and $\{c_i\}$ to an $N$-dimensional vector $\overline{C}$:

$$\overline{C} = \phi\overline{X}, \tag{4.75}$$

where $\phi$—square matrix $N \times N$ having the elements in the form

$$\{\phi\}_{in} = \frac{1}{\sqrt{N}} e^{-jn\frac{2\pi}{N}i}. \tag{4.76}$$

The inverse DFT is

$$\overline{X} = \Phi^{-1}\overline{C}, \tag{4.77}$$

where $\phi^{-1}$ is a matrix inverse to $\phi$, i.e., $\phi^{-1}\phi = 1$. The elements of matrix $\phi^{-1}$ are the following

$$\{\phi^{-1}\}_{in} = \frac{1}{\sqrt{N}} e^{jn\frac{2\pi}{N}i} \tag{4.78}$$

Thus, the elements of matrices (4.75) and (4.76) are of the form of the complex-conjugated quantities, while matrix $\Phi$ is unitary. Therefore, we may consider the DFT as a transformation of coordinates in the $N$-dimensional space using operator $\Phi$ that preserves the vector length, i.e.,

$$\|C\| = \|X\|. \tag{4.79}$$

This relation resents an analogue of Parseval's theorem. Rotation of the coordinates changes only representation of the vector that can be in this case either of the time or frequency kind. Most convenient is the representation enabling to diminish the number of non-zero components of the vector. Thus, in a time space a harmonic signal section has a great number of different from zero coordinates, whereas in the frequency space only one coordinate shows difference. As it was previously mentioned, the transfer from the orthogonal basis of trigonometric functions to special functions simplifies significantly the analysis of signals and elevates accuracy in a series of problems. The computer turns to be a most useful signal analyzer in this case.

A complete analogy between the DFT and signal filtration is improbable since a signal component remains unchangeable in the filter output, while the DFT transforms them from the time form into the frequency one. At the same time, a number of DFT parameters is analogous to the band-pass filter ones. The transmission band analogue of the filter is the DFT resolving capacity in frequency. As for the time of attenuation of transient processes in the filter, it coincides with that of a signal section being stored $\Delta T$.

The dependence of resolvability of a digital spectral analyzer upon a stored section duration is shown in Fig. 4.6. It illustrates a group of sections of a complex signal consisting of a sum of two harmonic signals with close frequencies and spectra of these sections [7, 9].

The roll off slope of the amplitude-frequency characteristic of the filter corresponds to the parameter of expanding spectral constituents of a signal DFT.

So, the spectrum of a harmonic signal section looks like

$$A(\omega) = A_0 \frac{\sin[(\omega - \omega_0)\Delta T/2]}{[(\omega - \omega_0)\Delta T/2]} \tag{4.80}$$

where $A_0$, $\omega_0$ and $\Delta T$ are the amplitude, frequency and time of the signal section, correspondingly.

The signal section acquires such a spectral form since it consists of a product of the harmonic signal multiplied by a function called a "rectangular time window". The spectrum of the product of two functions is known to be equal to the convolution of the function spectra, while the "window" function spectrum has a finite width. To reduce spectral density of a finite time signal, at frequencies, $|\omega - \omega_0| > 2\pi\Delta T$ it is worthwhile using time windows of a complex form, e.g., Hemming's or Hann's ones [1, 10].

The best results are achieved with the windows like $\frac{\sin x}{x}$, where $x = \pi(2t - \Delta T)/\Delta T$. However, this type of windows is hard to apply for the practical analysis of the noise and vibration signals due to a large volume of computation.

The accuracy of solving problems of deterministic signals on the background of random constituents using $C_n$ values derived from a fixed-duration sampling does not often correspond to a required level. So, one should either extend the sampling time with corresponding increase in resolvability of the analyzer and reduce interference proportionally to $\sqrt{\Delta T}$, increasing thereby computation volume, or accumulate the results of analysis of sequential signal sections, which results in a $\sqrt{m}$ times reduction of the interference value, where $m$ is the number of signal sections.

Averaging of the amplitudes of spectral components following the relation

$$\overline{A_n} = \frac{1}{m}\sum_{i=1}^{m} A_{ni} \tag{4.81}$$

is equivalent to the function of an ideal integrator in the analogue spectrum analyzer. To obtain the results corresponding to the operation of a real integrator, the amplitudes should be averaged as follows:

$$\overline{A_n} = \frac{1}{m}\sum_{i=1}^{l+m} A_{ni} - A_{nl}, \tag{4.82}$$

where $l$—number of the averaged period patterns.

The computer phase analysis of the deterministic vibration signals using the DFT has its peculiarities. One of them is the fact that the phase value of the harmonic component depends upon the discretization time start of a signal section $\Delta T$. In this connection, the difference between the initial phases of the frequency multiple vibration signal components found from the initial phases of the harmonic constituents acquires a physical sense. Another specific feature of the phase analysis is manifested when there occurs a frequency misalignment between the harmonic signal component and one of the frequencies $k\Delta f$ between the initial phase of the component and its measurement result leading to a systematic error proportional to the difference $\omega_i - 2\pi k\Delta f_p$, where $\omega_i$ is a circular frequency of the $i$-th component of the periodic signal.

To store the measurement results of phase characteristics, it is convenient to use the synchronous methods of the digital signal processing in the case the discretization frequency is proportional to those of the harmonic signal components. Such methods of signal analysis can be used when the computer is fit with additional information on the signal period. The synchronous methods of digital processing may use the synchronous signal to start from the ADC. This excludes systematic errors during phase measurements and raises efficiency of signal analysis when the vibration frequency is unstable.

The computation volume in determining coefficients of a signal section DFT is huge. It constitutes about $2N^2$ operations of multiplication and addition, where $N$ is the number of discretization moments. The DFT with a reduced volume of computations, called the fast Fourier transform (FFT), was put forward in 1965. At present, we know numerous modifications of the FFT with reduced by $\log_2 N$ number of operations, where $N$ is a memory span of the SU occupied by a signal section. Along with a multitude of the programs realizing the FFT for different computer types, there exist specialized microprocessors exercising the FFT at the circuitry level [9].

## References


1. R.B. Randal, Frequency Analysis. Brüel & Kjær Theory and Application Handbook BT 0007-11 (Brüel & Kjær, Nærum, 1987), p. 344
2. D.D. Klovskii, *The Theory of Signal Transfer* (Svyaz, Moscow, 1973), p. 376
3. Y.I. Iorish, *Vibrometry. Measurement of Vibration and Impact. General Theory, Methods and Instrumentatio* (Mashinostroenie, Moscow, 1963), p. 772
4. L.S. Pontryagin et al., *Mathematical Theory of Optimal Processes* (Fizmatgiz, Moscow, 1961), p. 391
5. A.A. Kharkevich, *Spectra and their Analysis* (Mashinostroenie, Fizmatgiz, Moscow, 1960), p. 392
6. M.K. Sidorenko, *Vibrometry of Gas-Turbine Engines* (Mashinostroenie, Moscow, 1968), p. 224
7. A.A. Belousov, *Diagnostics of Mechanical Systems of the Audiovisual Aids* (SPb Politekhnika, St. Petersburg, 2002), p. 152
8. S.S. Dobrynin, M.S. Feldman, G.N. Firsov, *Methods of PC-Aided Research of Vibration in Machine* (Mashinostroenie, Moscow, 1987), p. 224
9. Systems of information storage and processing, of Pulse series [Electronic resourse]. Moscow Techn. Center Brüel & Kjær (2011). http://bruel.ru/UserFiles/File/What_is_PULSE_clear_vers.pdf. Accessed 18 Dec 2011
10. L. Rabiner, *Theory and Application of Digital Signal Processing* (Prentice Hall, Upper Saddle River, 1975), p. 762

# Chapter 5
# Friction-Excited Self-oscillations

Friction-excited self-oscillations should be accounted for as an eventual source of vibrations in machines and mechanisms. This is one of the most spread types of mechanical self-oscillations connected with instability of the friction process under relatively low sliding velocities. Named phenomena are becoming typical for modern engineering, especially for the processes of fine positioning, starting and braking of the joints in various devices. The analysis of investigation results of friction-excited self-oscillations is presented, the conditions of their origination and untwisting are shown. The basic role of the staticokinetic characteristic of the rubbing materials in arising frictional instability under low sliding velocities is underlined. The mechanisms of exciting self-oscillations in the metal—friction composite tribopairs are considered. The role of damping in the appearance of the HF vibroacoustic activity in the tribopairs is stated. The design methods of friction-induced self-oscillations are analyzed for the macrosystems.

## 5.1 Self-oscillations in Friction Systems

When speaking about a friction system, we usually imply the rubbing components with a set of interrelated components influencing operation of the whole system. A set of components connected with the rubbing element forms a mechanical subsystem characterized by a certain stiffness and frequency of its eigenmodes. Depending on friction conditions and parameters of mechanical subsystems, one of which may turn to be more sensitive to oscillations, we accept parameters of the latter as the ones characterizing the whole frictional system [1].

During cutting, oscillations are excited in subsystems "cutter–support" relative to the in-process workpiece, and the "workpiece–machine frame" relative to the cutter. The mechanical parameters of a frictional system at high-velocity cutting depend on the cutter–support subsystem parameters, in the low-velocity cutting on the workpiece–machine frame ones.

A frictional system like any other mechanical system may induce various by their nature self-oscillations, e.g., the ones induced by a misbalance, eccentricity of



V.P. Sergienko and S.N. Bukharov, *Noise and Vibration in Friction Systems*,
Springer Series in Materials Science 212, DOI 10.1007/978-3-319-11334-0_5

the axles, or other. More often we encounter in such systems friction-induced self-oscillations (FS), i.e., the ones arising from instability of the friction process [1–3].

Discreteness of the actual contact creates a situation when the real friction process at the microlevel is persistently unstable. Multiple elastic and inelastic pulse deformations of microasperities, surface areas and microvolumes of surface layers together with competing wear and regeneration processes of surface films result in a wide-range micro-oscillations of contact surfaces.

The frequencies of these micro-oscillations are commensurable to eigenmodes of the mechanical friction system and may interfere oscillating processes in the friction zone. The amplitude of micro-oscillations may increase under certain conditions, thus making friction at the microlevel unstable. Instability of this kind is a source of energy for sustained FS.

The resonant (cavity) oscillations under which eigenmode frequencies of structural elements coincide with the driving forces in the FS source are considered as most dangerous ones. The resonant oscillations are the main source of vibrations and acoustic noise generated by the friction joints. The resonance amplitude in the systems devoid of oscillation damping may theoretically grow infinitely. In practice, the oscillation energy is being scattered in the system persistently. The energy of the external source of excitations is limited, while the structural elements have dissipative properties, wherefore the oscillation amplitudes are growing till some limit or until some weakest element of the system fails [4, 5].

Brakes and friction clutches are the examples of commonly used friction joints in which self-oscillations are excited mostly when sliding is nonuniform [6, 7].

The engaged friction clutch excites the FS in a transient period of acceleration of the driven half-coupling till the rotation velocity of the driving half-coupling. As for the mechanical automotive transmissions connecting the engine shaft with the driving wheels, the FS are accompanied during skidding by jerks of 8–10 Hz frequency [8] and a high-frequency noise [9].

Vibration of the airplane brakes caused by friction-induced self-oscillations is inadmissible in practice as it may destroy not only the brake itself but the wheels and landing gear elements too [10]. The FS in the systems of indirect control are generated by friction in servomotors, especially in the ones controlled by a slide valve of the flapper-nozzle type or a jet pipe [11, 12]. FS-induced cyclic shear in most of above-mentioned friction joints may lead to fretting and catastrophic wear [13] accompanied by the noise [14]. From the other side, friction-induced acoustic oscillations present an additional effective loading on the tribosystem that intensifies wear processes in the rubbing pair [15].

The FS may impart judder to a slow feed in almost all types of machine-tools: turning, milling, boring, etc. This is especially unacceptable in PC-aided high-precision machines. In some cases, to avoid cost rise of the drives they accept unevenness of the slowest feed to regulate admissible judder. The FS may also impair sensitivity of the feeding mechanism of accurate adjustment, e.g., when centering holes at boring [16]. The resulted error in positioning may surpass half of the machining error [17]. To meet the precise dimensions the error in positioning heavy-duty turning and milling machines may constitute 10 μm [18].

The FS may occur in dynamically locking self-stopping jacks, self-braking angular positioning mechanisms of heavy astronomic tools, heavy instruments, manipulators, and so on [19].

Friction and nonlinear resistance to plastic deformations during wire drawing generate relaxation oscillations, which are especially undesirable at drawing high-strength wire. Along with worsened lubrication conditions and mechanical properties, they may lead to distortion of geometry by forming cross rings and longitudinal undulation of the wire [20].

It is evident that self-oscillations in friction systems belong to a critical type of mechanical oscillations that restrict accuracy and reliability of machines and mechanisms. Therefore, the problem of their thorough study stands on the forefront of investigations in this field.

## 5.2 Investigations of Friction-Induced Self-oscillations

There are two chief approaches to investigations of frictional self-oscillations, namely, mathematical and physical methods [1].

The mathematical approach treats the idealized tribological properties by deducing and solving differential equations of motion proceeding from the notions of inertia, velocity, rigidity and damping. The equations of motion of the friction systems are obtained in the first approximation rather trivially. Analytical mechanics has not as yet elaborated a universal method for the systems like, e.g., Lagrangian one to describe frictionless systems. Therefore, dynamics is analyzed in practice only for a certain class of friction systems or their operation mode, e.g., vibration-induced displacements [21]. The problem of how to calculate accurately the FS in the real mechanisms with the known kinematics and motion parameters still remains unsolved today. This is, in particular because of scarce scientific data on the mechanisms of FS generation.

The physical approach considers a simple oscillating system (commonly by neglecting dynamic characteristics of the design), and studies in detail tribological aspects of FS. The main of them are the dependencies of friction parameters on the static and kinetic characteristics of the rubbing bodies, as well as the properties of structural and lubricating materials (LM) of the complete elastic system.

It is evident that interpenetration and complementary character of these two approaches in the research and development of the new means of preventing FS in tribounits forms a specific science on the borders of the theories of oscillations, dynamics of machines and tribology [1].

Investigations of FS can be subdivided into two groups depending on the purpose [13, 21].

(1) Evaluation of a preset motion stability for all systems in which FS are inadmissible. The conditions of stability are commonly set forth as the critical values (permissible minimum) of sliding velocity, rigidity and damping in the system;

(2) Evaluation of FS parameters for the systems where FS are expedient (e.g., constitute a working function) or admissible under the least velocity value (esp. seldom used). In the latter case the permissible maximum amplitude of self-oscillations is preset.

A peculiarity of FS—so-called stick-slip effect implies that the oscillating process consists of two unlike in character stages of a stoppage and a leap, i.e., of a uniform motion at a relative macrorest of the rubbing bodies with a probable microslip, and further nonuniform jerking relative displacement [1]. The diagrams of the FS are ranging depending on the relative duration of these stages from a sawtooth till sinusoidal forms. When the velocity comes periodically to zero the FS are termed as relaxational. It is proposed in [22] to subdivide them into two classes: discontinuous and continuous FS (with/without stops).

The discontinuous motion proceeds as follows. Initially, the elements of the friction system undergo deformation under the external forces till the force transmitted by them surpasses the total force of the static friction so as to begin sliding. As the friction force drops, the motion accelerates and then stops, terminating the leap. The deformation of structural elements tends to diminish after the leap, so that to continue the motion some period of stoppage is needed, during which the deformation increases again. This is followed by a next leap, wherefore we can speak about discontinuous oscillations.

The authors of [23, 24] have stated that self-oscillations with a periodic stoppage occur when the static friction exceeds the dynamic one.

One of the first mechanical models of FS consisting of a weight moving along a rubber band was proposed by Den Hartog [25] and Van-der-Paul [18]. The presence of the friction force imparting nonlinearity to the system with one degree of freedom was considered as a condition for generation of relaxation oscillations.

A strict mathematical description of the mechanical self-oscillations of the first kind occurring at dry friction was first presented in [26]. The general theory of oscillations was on the main applied to electrical oscillations. But this theory can be extended for the mechanical oscillations too, as it was shown by N.L. Kaidanovskii and S.E. Haikin.

Self-oscillations in mechanical systems, at least in those with one degree of freedom, are probable when the linear friction characteristic makes provision for the nonlinear nonconservative property of these systems. Since in a general case the friction force is a nonlinear function of velocity, all friction systems possess a nonlinear non-conservatism. To excite self-oscillations, this non-conservatism should be well defined, e.g., in some fields the kinetic characteristic of friction should be falling. In practice, this fact is often observed in the field of low slifing velocities in conditions of dry and starvation friction.

The equation of motion for a system with one degree of freedom is of the type:

$$m\ddot{x} + cx = F(\dot{x}) \tag{5.1}$$

where $F$—friction force as a function of velocity $\dot{x}$.

Similar to the electric self-oscillating circuits, where we discriminate between Thomson's and relaxation self-oscillations that are dependent on the role of electric resistance in the system, N.L. Kaidanovskii and S.E. Haikin have examined two limiting cases differed by the role of friction in the system:

(1) Friction force variations are negligible and their role is limited only to self-generation of oscillations, where friction defines the amplitude of stationary oscillations but does not in fact influence their frequency and type, which is close to sinusoidal one. The oscillation frequency is conditioned by the inertial force (mass, inertia moment) and the restoring force (elasticity, gravity force, and so on) that are roughly equal to one another:

$$m\ddot{x} + cx \approx 0 \tag{5.2}$$

(2) Variations in the friction force are so strong that the period and type of oscillations differ much from those of harmonic oscillations without friction, the oscillation frequency becomes lower and the form is sinusoidal, while the inertial force and the restoring force may differ several-fold or even by orders of magnitude.

In the first case, the period and form of oscillations are roughly similar to a system not frictional at all, i.e., their frequency approaches eigenfrequency of the system. This case has been studied as applied to the Froude pendulum [27], which is suspended to allow friction on a uniformly rotating shaft. Small oscillations of the pendulum are described by the equation:

$$I\ddot{\varphi} + M_{н}(\varphi) + M_{в}(\dot{\varphi}) + M_{m}(\dot{\varphi}, \omega_{в}) = 0, \tag{5.3}$$

where $M_{н}$—moment of gravity force dependent on deviation angle of the pendulum $\varphi$; $M_{в}$—moment of air resistance force being a function of $\varphi$; $M_m$—moment of friction force dependent on $\varphi$ and angular velocity of the shaft $\omega_{в}$. Solution of (5.3) gives us a relation for the amplitude of quasi-harmonic oscillations of the pendulum and four probable types of behavior of the system:

slight excitement of oscillations and a limiting cycle;
damped oscillations;
increasing oscillation amplitude till infinitely large values per finite time;
unstable limiting cycle.

The condition of harmonic self-oscillations is presented in [27] as follows:

$$-\frac{dF}{d\vartheta} \geq k_c, \tag{5.4}$$

where $\vartheta$—velocity, $k_c$—resistivity factor of medium.

It is difficult to estimate the second case in mathematical terms, especially when the oscillations are non-sinusoidal. Therefore, simplifying suppositions are required, e.g., the one on a negligibly small mass. As a result, the oscillating process decomposes into two different regions:

(1) the elastic force exceeds much the inertial force, so it follows from (5.2)

$$cx \approx F(\dot{x}). \tag{5.5}$$

Due to considerable changes in the coordinate in this region, acceleration is insignificant and the velocity is varying slowly, while elasticity undergoes noticeable changes;

(2) the inertia force is larger than the elasticity force variable, so proceeding from (5.2):

$$m\ddot{x} \approx F(\dot{x}). \tag{5.6}$$

The coordinate of the system changes but little in this region, while the velocity alters much due to significant accelerations.

The time of passing the second region is negligibly small as compared to the first region, so far it is suffice to study only the first region, in provision that the oscillating system mass is insignificant. The oscillating process in this system consists of four stages: two regions where the system is moving uninterruptedly with continuously varying velocity, and two positions in which the velocity varies by leaps under invariable coordinate, i.e., the relaxation oscillations are taking place.

During the experimental check of the model [26], the present authors have established the data, which is inexplicable within the frames of accepted assumptions:

- the effect of the "retarding" self-excitement, i.e., the oscillations appear under the lower velocities than they terminate (the probable causes lie in the peculiarities of function $F(\dot{x})$ in the region of minimum or in unaccounted surplus degrees of freedom);
- the amplitude of the first oscillation surpasses noticeably the next ones (because of the involved statistical characteristic of friction to be discussed below).

In spite of some drawbacks, the model of discontinuous mechanical oscillations [26] can be applied in the first approximation to describe non-sinusoidal oscillations in the systems experiencing intensive friction. In his special experiment, N.L. Kaidanovskkii has shown that self-oscillations under the velocities complying with the falling friction characteristic can be removed only when the critical damping factor in the drive surpasses the slope factor.

The critical damping factor was determined on the base of the viscous friction, and the slope factor by finding the slope angle tangent of he tangent line to the curve of the kinetic characteristic of friction.

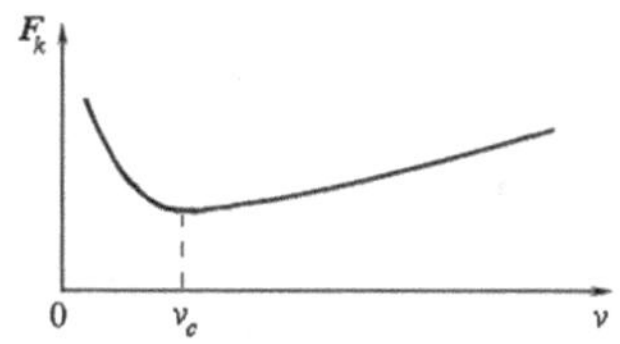


**Fig. 5.1** Kinetic characteristic of friction

The presence of critical velocity $v_c$ above which self-oscillations disappear can be explained by transition from the falling to growing frictional section of the friction characteristic (Fig. 5.1). It should be noted that the presence of the critical velocity separating the regions of smooth and discontinuous motion is an important property of the FS.

A general solution of this problem for self-oscillations was derived some time later in work [28] where the friction force was presented as consisting of two parts, namely, an ideal Coulomb's one and a linearly viscous part (Chap. 2, Table 2.2).

According to [26], there are two limiting cases, namely, harmonic oscillations under a low friction force and relaxation oscillations under a high friction force, which are particular cases of the general solution.

It is considered in works [29–31] that the falling type of the kinetic characteristic of friction at boundary and mixed lubrication modes is the reason of the FS. It is shown in [30] that harmonic self-oscillations are observed at $dF/dv \approx 0$, which attenuate at slight damping. The relaxation self-oscillations at $dF/dv \ll 0$ are feasible when the system is insufficiently rigid, whereas in the case of its high rigidity, the harmonic self-oscillations are more probable. The parameters of the FS depend also on LM properties, particularly, on their dynamic viscosity [32], and the oscillation frequency grows with sliding velocity.

In the case of dry friction of steel surfaces, the frequency of self-oscillations $\omega$ in the region of high loads increases as the pressure $P$ and velocity $v$ grow. This dependence is satisfactorily described by equations of the type [33]:

$$\omega = k_1 e^{k_2 P}, \tag{5.7}$$

$$\omega = k_3 v^{k_4}, \tag{5.8}$$

where $k_1 \ldots k_4$—positive empiric coefficients; $k_1$ and $k_2$ are dependent on sliding velocity, $k_3$ and $k_4$—on contact pressure, which growth reduces $k_1$ and increases $k_2$.

So, path $S$ passed per self-oscillation period and defined by relation $\omega/v$ can be expressed by the formulas:

$$S = \frac{v}{k_1} e^{-k_2 P}, \tag{5.9}$$

$$S = \frac{v}{k_3} v^{1-k_4}. \tag{5.10}$$

Diagrams $\omega(v)$ made at different pressures in the logarithmic scale look like straight lines intersecting in the point with abscissa equal to critical velocity $v_c$, while the ordinate is roughly equal to eigenfrequency $\omega_0$ of the system. It is stated in [34] that graphs $\omega(v)$ are not connected with rigidity of the system but depend upon tribological properties of rubbing materials. It is also noted in works [35–37] that the dependence of the critical velocity on frictional characteristics seems more justified.

Thorough experiments with different friction pair materials were performed in work [38]. As a result of investigations, the following regularities were established:

strict periodicity between leaps and stops;
presence of the first and second critical velocities that determine transition from the continuous to discontinuous motion and back, i.e., thus limiting the region of the stick-slip motion;
proximity of the leap duration to half-period of eigenmodes of the system
increase of stoppage time with reducing mean sliding velocity.

A peculiar feature of self-oscillations is a strong difference between the maximal and minimal velocities during the leaps. As far back as in 1939, F. Bowden and L. Leben have put forward a theory that explains the stick-slip motion of the rubbing bodies by failure of the weld bridges in the contact zones [39]. In the course of investigations of steel tribopairs on a low-rigidity laboratory facility at dry friction with a feed rate 0.06 mm/s, they have observed about 100–1,000 mm/s maximal velocity, and the minimal one equal to zero. It is of interest that under the high loads, Jones [40] has recorded the mean leap rate surpassing the nominal one by as much as 1,500 times.

Nevertheless, the relaxation oscillations are typical also for the pairs like wood—steel, cast iron—polymer frictional materials, that exclude the probability of welding owing to their nature. Besides, the stick-slip motion can be observed under such loads that are insufficient for the weld bridge to form. In this connection, Bowden's theory turns to be true only for the cases of the molecular seizure of surfaces.

The decisive role in exciting FS in work [41] is attributed to discontinuity of plastic deformations in localized volumes of the elastically fixed element of the friction pair. It was also stated that the number of oscillations corresponds always to that of the discontinuous cohesive transfer traces on the friction tracks of a softer material. The results of numerous investigations e.g., [42] have proved an important role of transfer films (third body) in altering adhesion between contacting surfaces and, consequently, in varying the parameters of self-oscillations. Mentioned phenomena are illustrated in Fig. 5.2, where the transfer films are indicated by arrows.

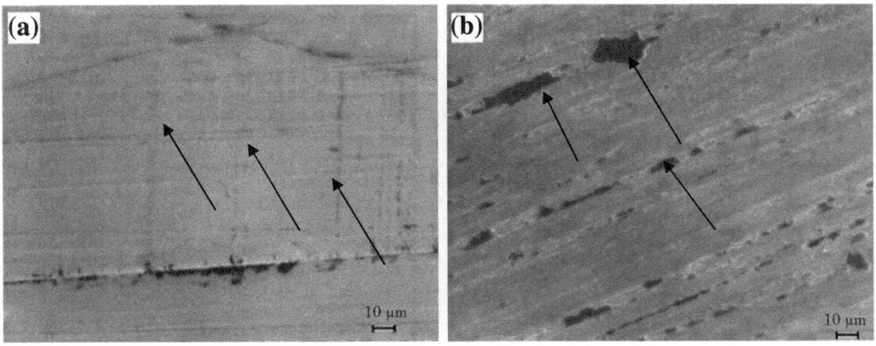


**Fig. 5.2** SEM images of steel counterbody surface after rubbing against a polymer material: **a** cohesive traces of polymer transfer formed at the initial stage of sliding, **b** transfer films (third body) [43]

## 5.3 Statico-Kinetic Characteristics of Friction

It is evident from the above procedure that the oscillating process in a system in the presence of FS consists of different in principle stages. During the first stage the rubbing surfaces are found in a relative rest that depends on operation conditions and *static characteristics* of the friction pair. In the second stage the rubbing surfaces are in condition of relative motion, which lasts depending on operation conditions and *kinetic characteristics* of the tribopair. These stages in combination are determining the period of friction-induced oscillation. Therefore, in studying friction oscillations one should account for not only the friction coefficient value (level) but also the kind of kinetic and static characteristics in certain operation conditions since each of them has its specifics because of differences in friction-induced interactions [1]. The theory of Kosterin and Kragelskii [34] states that the tribological conditions of frictional self-oscillations can be described in a general case by a system of equations:

$$\begin{cases} \dfrac{dF_k}{dv} < 0 \\ \dfrac{dF_s}{dt_n} > 0, \end{cases} \tag{5.11}$$

where $F_k$ and $F_s$—kinetic and static friction forces, correspondingly, $v$—sliding velocity, $t_n$—stationary contact duration.

In other words, the condition for the FS to be excited in a system is a falling kinetic $\frac{dF_k}{dv}$ and rising static $\frac{dF_s}{dt_n}$ frictional characteristics.

The theories explaining excitement of the FS by differences in friction forces of the rest and sliding [44, 45], i.e.,

$$\Delta F = F_s - F_k > 0 \tag{5.12}$$

interpret a simplified version of above-named theory by Kosterin and Kragelskii [34].

The statico-kinetic characteristics of friction are dependent on the strain state in the rubbing bodies, lubrication type or its presence, material properties of the friction elements and many other parameters of the friction system. It is natural that these parameters are varying in different time scales during operation, therefore, the FS may be either excited or damped.

### *5.3.1 Kinetic Characteristic of Friction*

In a general case, the kinetic characteristic of friction can be presented by the known Stribeck diagram that sets the friction coefficient as a function of a complex parameter $\frac{\eta\omega}{p}$, where η—dynamic viscosity, ω—angular rotation velocity of the shaft, *p*—contact pressure (Fig. 5.3).

The first piece of this curve characterizes dry friction with a high friction coefficient. Transfer to the boundary friction is accompanied by the intensive reduction of the friction coefficient followed by a piece of mixed lubrication, where the intensity of falling retards, and after passing a minimum on the piece of hydrodynamic (fluid) lubrication it starts growing. The typical mean values of the friction coefficients at sliding for different lubrication modes are shown in Fig. 5.1.

An attempt to analyze the kinetic characteristics for metal sliding without lubrication was made in work [46] proceeding from the phenomenon of plastic flow

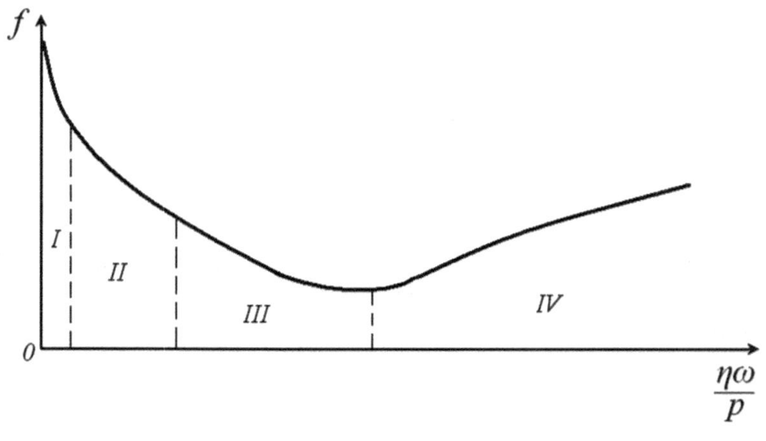


**Fig. 5.3** Stribeck diagram: *I*—dry friction; *II–IV*—friction in boundary, mixed and fluid lubrication conditions, correspondingly

**Table 5.1** Variation range of sliding friction coefficient

| Friction mode | Friction coefficient |
|---|---|
| Dry friction: | |
| in vacuum | ≥1.00 |
| in atmosphere | 0.1–1.00 |
| Boundary friction | 0.050–0.250 |
| Mixed friction | 0.010–0.050 |
| Hydrodynamic friction | 0.001–0.010 |

in the friction contact zone (Table 5.1). The friction force was taken proportional to the actual contact area (ACA) $A_r$

$$F = \tau A_r \tag{5.13}$$

where τ—specific maximal shear stress in the contact zone. Metal flow proceeds at a variable speed depending on the ACA, hardening and recovery processes of the metal. In the case the contact occurs quickly, the ACA is small and the velocity $dA_r/dt$ is high. If the time of contact increases, the ACA enlarges till some maximal size $A_r^{\max}$:

$$\frac{dA_r}{dt} = -k(A_r - A_r^{\max}), \tag{5.14}$$

where $k$—a constant characterizing properties of materials. Under the elevated sliding velocities metal surfaces in the friction zone are more hardened due to intensive surface deformations:

$$t_0 = \frac{l_0}{\vartheta}, \tag{5.15}$$

where $t_0$—mean time of metal flow between two sequential microcontacts, $l_0$—mean distance between microcontacts. If to substitute (5.15) in (5.14), we obtain a differential equation:

$$\frac{1}{A_r - A_r^{\max}} dA_r = \frac{kl_0}{\vartheta^2} d\vartheta. \tag{5.16}$$

Integration of (5.16) gives

$$A_r = A_r^{\max} - (A_r^{\max} - A_r^{\min}) e^{-\frac{kl_0}{\vartheta}} \tag{5.17}$$

where $A_r^{\min}$—minimal ACA at $\vartheta \to \infty$. So, from (5.13) we may obtain a relation for the kinetic characteristic of friction

$$F_k = F_0\left(1 - k_a e^{-\frac{kl_0}{\vartheta}}\right), \tag{5.18}$$

where $k_a = \left(A_r^{\max} - A_r^{\min}\right)/A_r^{\max}$—coefficient characterizing plastic flow at compression.

It is commonly accepted that the friction force in dry friction is independent of sliding velocity. In practice, this is true only for the case when the temperature in the contact zone varies but insignificantly and, consequently, the surface layer properties do not change [47]. Therefore, the falling dependence of the friction coefficient versus sliding velocity (kinetic characteristic) in dry friction was observed by a number of researchers starting from Rayleigh [48].

The kinetic characteristics of friction of polymers and their composite materials are dependent on contact pressure. The declining kinetic characteristics of friction were obtained in work [49] for polycaprolactam–steel rubbing under high pressures and the rising ones under the low pressures. The inverse dependencies were derived for textolite. This is, probably, connected with oxidation processes and formation of the films of secondary structures. Named characteristics are rising at insufficient or excess intensity of above-named processes, and are falling at some lower intensity [1].

The dependence of kinetic characteristics on the contact pressure during friction in vacuum was also discussed in [50] for the samples of hardened steel U10 and steel 45 under 6–60 cm/s velocities.

In the ultralow velocity range (about a micrometer per minute) the kinetic characteristics for the steel—indium and steel—lead pairs without lubrication tend to increase [51]. The increasing kinetic characteristics under ultralow velocities were also obtained in works [52, 53] for the friction joints of transport vehicles. It was noted that the friction force dropped with reducing velocity almost linearly and became close to null at a zero speed.

The kinetic characteristics for friction without lubrication at an iron disk sliding against a copper or tin cylinder with 0–145 cm/s velocity were obtained in work [46]. The experiments were conducted under −180 till +210 °C. It was shown that under low temperatures the kinetic dependencies were more flat in contrast to higher temperatures. The analogous results were obtained for a series of frictional materials by Kosterin in [54].

Approximation of the kinetic characteristics of friction using a quadratic or cubic parabola, exponent, or some other continuous monotonous function [14, 16, 55, 56] has brought us to solution of the differential equation of motion in a closed analytical form. As for the piecewise-linear approximation, it does not always a necessary an analytical solution [16, 57].

It is rather difficult to analyze in practice the effect of sliding velocity and contact temperature separately. Therefore, one should thoroughly interpret the data selected from kinetic characteristics presented by different researchers. Table 5.2 lists some results for tribopair elements made of various polymeric materials [47].

Mixed lubrication is very often applied in practice, in particular, it is characteristic for control mechanisms of machine-tools. One of the first theoretical investigations of kinetic characteristics in mixed lubrication was performed in [66], where the normal and tangential contact forces were determined as a sum

**Table 5.2** Kinetic characteristics of polymer materials

| Number | Author | Material | Sliding velocity | Graphical representation |
|---|---|---|---|---|
| 1. | Shooter and Thomas [58] | PTFE, PE, PMMA, PC | 0.01–1.00 cm/s Steel–polymer | 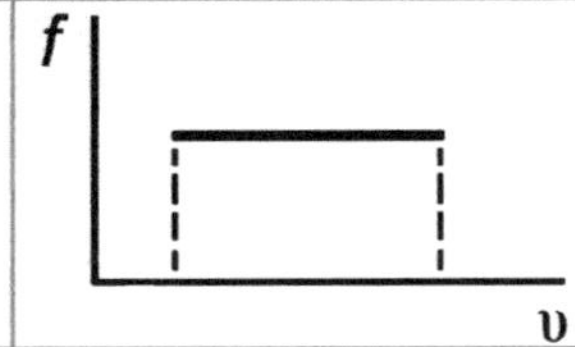 |
| 2. | Milz and Sargent [59] | 1—nylon; 2—PC | 4–183 cm/s Polymer–polymer | 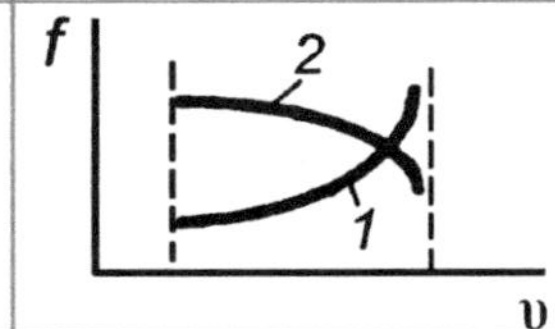 |
| 3. | Fort [60] | PTFE | $10^{-5}$–10 cm/s Steel–polymer | 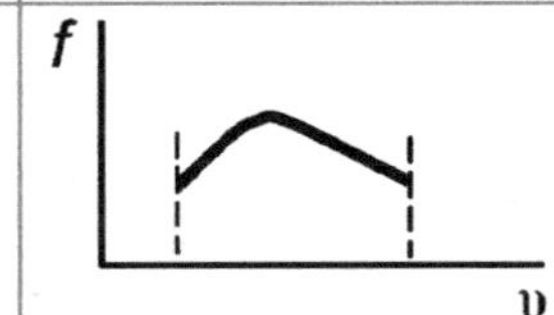 |
| 4. | White [61] | 1—PTFE; 2—nylon | 0.1–10.0 cm/s Steel–polymer | 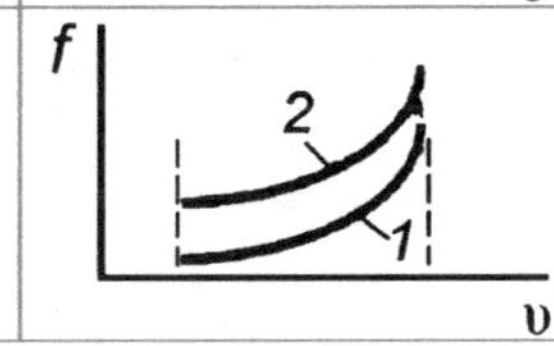 |

(continued)

**Table 5.2** (continued)

| Number | Author | Material | Sliding velocity | Graphical representation |
|---|---|---|---|---|
| 5. | Flom and Porile [62, 63] | PTFE | 1.1–180.0 cm/s<br>Steel–polymer | 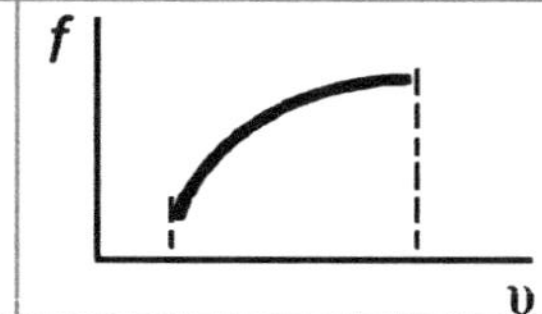 |
| 6. | Bartenev and Lavrentiev [64], Schallamach [65] | Rubber | Design | 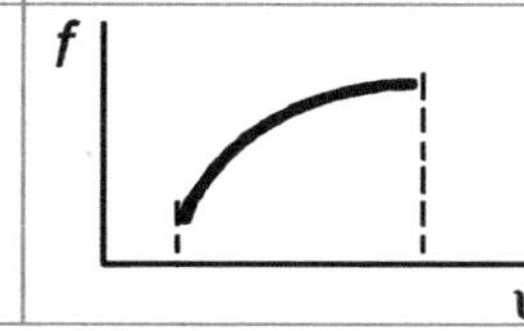 |

of interactions between surface asperities and hydrodynamics of a viscous fluid. This is expressed in the formula:

$$f_{mix} = f_{dry} - k\frac{\mu\vartheta}{N}, \tag{5.19}$$

where $f_{mix}$ and $f_{dry}$—friction coefficients for the mixed and dry friction, correspondingly; $k$—empirically determined coefficient.

The linear dependence (5.13) based on the assumption that the minimal clearance between the rubbing surfaces is constant and equal to the sum of microasperity heights of contacting surfaces seems to be rather approximate.

Summation of molecular forces between surfaces and the forces of viscous resistance in the lubricating layer was accepted in works [67, 68]:

$$f_{mix} = k_1\vartheta^{\frac{1}{2}} + k_2\vartheta^{-1}, \tag{5.20}$$

where $k_1$ and $k_2$—coefficients dependent on the friction conditions. In the case of the vanishing velocity, formula (5.20) gives an infinitely large value of $f_{mix}$, thus limiting its application.

Friction at mixed lubrication representated in [8] as a relationship between solids, viscous resistance of the fluid and its hydrodynamic effect allowing for deformability of the bodies under the contact loading. This concept has been further developed recently within the frames of the hydrodynamic theory of mixed lubrication by Kudinov [16], Birchall and Moore [69]. It is common knowledge that macro and microcavities found between the actual contact areas are interconnected and filled by the initial and waste LM, wear debris, and etc. These cavities have the portions of narrowing and broadening through their height. The intermediate medium together with the sliding surface are involved into a relative displacement forming thereby a hydrodynamic wedge in the places of narrowing. The summarized effect of these macro- and microwedges may result in elevation of one surface over the other. The appearance of the hydrodynamic effect on microasperities of the actual surfaces was supported by a number of researchers [70].

According to Kudinov's theory, the lifting capacity $Q$ equals to a sum of lifting forces of the microwedges formed in the contact:

$$Q = \frac{6\mu k_g}{tg^2\alpha}, \tag{5.21}$$

where $k_g$—wedge length ratio, $\alpha$—mean inclination angle of wedges.

As the lifting increases, the contact deformation and a part of the normal load perceived by the surfaces reduce, while the part of the normal load born by the lubricating layer increases. As the velocity increases from zero the friction force reduces since a large share of the load is perceived by the lubricating layer. In some point of the kinetic curve the friction reaches its minimum in the case the surfaces are fully separated by the lubricating layer approximately equal to the rough layer thickness. As the velocity continues to increase, the friction force augments in accordance with hydrodynamic laws.

Some time ago we believed that the falling kinetic characteristic is requisite for the frictional self-oscillations to appear. Kaidanovskii [26, 27, 71], Schnurmann and Warlow-Davies [31] and some other scientists have proved that this fact is true also in conditions of boundary lubrication by obtaining an experimental kinetic characteristic corresponding to the friction curve shown in Fig. 5.3. The falling section of this characteristic is explained in work [30] by a nonlinear behavior of the electrostatic component of the friction force, which is significant if the boundary layers separating the surfaces display either dielectric or semi-conducting properties. The phenomena of electrostatic discharge are closely related with relaxation oscillations. They consist of sequential alternating cycles of a slow charging during which a work on overcoming the field of unlike electrostatic charges on the conjugated surfaces is done, and the cycles of a fast discharging initiated after separation of the charges prior to a breakdown voltage on the boundary film, transforming it from the dielectric into a semiconducting one. These cycles comply with the ones of a slow microslip (stop) and fast sliding (jump) at a velocity below $\vartheta_k$. The experimental works [30] have shown that the jumps coincide with the electric charge of the galvanometer connected to the conjugated elements, and the value of these jumps reduces as the dielectric breakdown voltage on the lubricating film drops.

According to Tolstoy and Biny-Yao [72], so-called "instantaneous leap" of the friction force appears at stoppage due to a strong steepness of the falling kinetic characteristic of friction, and also because the factors inducing the static friction growth some time after the stoppage start to appear just before the full stop.

The abrupt drops of the instantaneous values of sliding velocity during FS are ranging from about zero till the maximal ones surpassing about tenfold the nominal velocity. The corresponding variations in contact deformation and the state of surface films governing the friction force lag behind the fast velocity changes (accelerations) due to certain inertia. So far, the friction force during FS changes slower than the instantaneous sliding velocity. The magnitude of the friction force changes less than the difference between the friction forces of the stationary sliding effects with velocities equal to the maximal and minimal values of the instantaneous velocity of self-excited oscillations.

At a nonsteady motion we should differentiate between the *kinetic* and *dynamic* dependencies of friction. The latter is found under considerable accelerations and continuous velocity variation, while the former occurs at a stepwise velocity change. The difference of the kinetic and dynamic frictional characteristics depends upon the degree of velocity non-uniformity [31, 36, 73–76].

Rabinovich [77] has proved that the instantaneous friction coefficient at a fast velocity variation depends upon the mean sliding velocity on the preceding section of the path equal to $10^{-5}$ m and roughly corresponding to a mean size of the actual contact unit sites. At instantaneous acceleration till a given velocity the friction force reaches a stable value equal to this velocity after a $10^{-5}$ m path. In the case the velocity is accelerated in a pulse from the rest, the friction force of rest would not change over a path about $2 \times 10^{-6}$ m, and reduces after a $10^{-5}$ m path till the friction force of sliding. Such a "memory effect" can be an explanation to the experimental data obtained in [78] and presented in Fig. 5.4. The friction coefficient

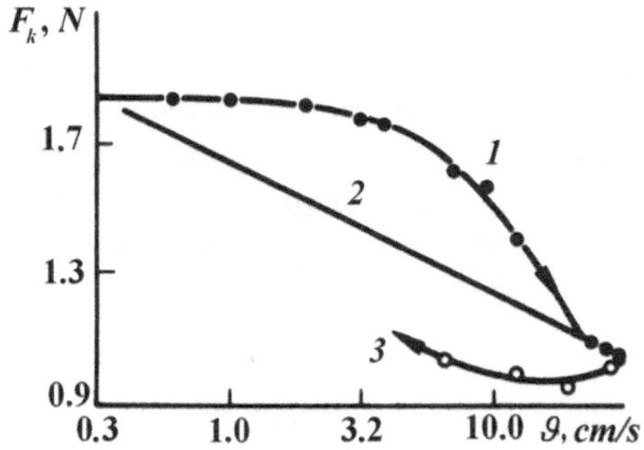


**Fig. 5.4** Dynamic (*1*, *3*) and kinetic (*2*) characteristics of friction

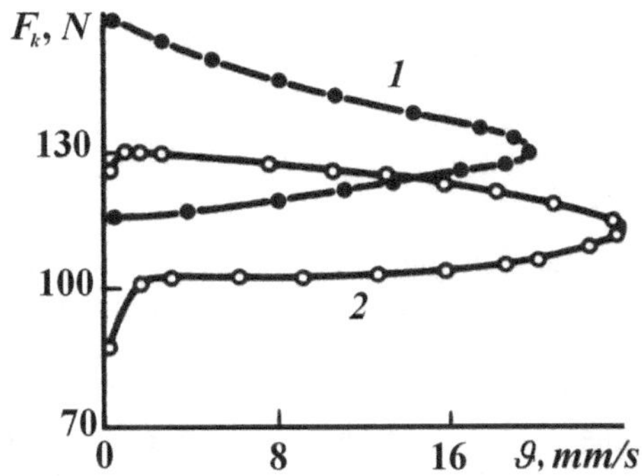


**Fig. 5.5** Two-valued dynamic characteristics of friction: *1* and *2*—different values of $Q$ [76]

is higher at acceleration (curve 1) and is lower at deceleration n contrast to the uniform motion (curve 2). This is because the instantaneous friction force values are defined by the velocity on the preceding $10^{-5}$ m long paths that is lower at acceleration and higher at deceleration as compared to above-considered moments. The dynamic characteristics of friction with varied velocity are often presented as a set of two-valued curves shown in Fig. 5.5 [79]. This, however, complicates experimental determination of the kinetic characteristic at low sliding velocities. In this connection, it is required either to add stiffness to the test bench drive or include special dampers in the mechanical system [80].

The friction force under a uniform motion can be determined experimentally from the elastic force in the drive $F_y$

$$F = F_y, \tag{5.22}$$

Under a non-uniform motion one should also take into account the inertial forces $F_u$ and those of damping $F_{\partial}$:

$$F = F_y + F_u + F_{\partial}, \tag{5.23}$$

i.e., it both the displacement, and acceleration (velocity) should be taken into account. When we determine, e.g., the dynamic characteristic of friction using the stick-slip oscillograms, the given velocity $\dot{x} = \vartheta$ is observed only in the oscillogram extremums, while in other points it is $\dot{x} \neq \vartheta$. According to [80], the friction coefficient $f_{kv}$ corresponding to this velocity is

in the maxima with the ordinate:

$$f_{kv} = cx_{\max} - m|\ddot{x}_{\max}|, \tag{5.24}$$

in the minima with the ordinate $x_{\min}$:

$$f_{kv} = cx_{\min} + m|\ddot{x}_{\min}|, \tag{5.25}$$

Consequently,

$$f_{kv} = c\frac{x_{\max} + x_{\min}}{2} - m|\ddot{x}_{\max} - \ddot{x}_{\min}|. \tag{5.26}$$

The dynamic frictional characteristics were studied in [75, 79, 81] and other works by a simultaneous measurement of displacements, velocities and accelerations. The investigations have shown that the dynamic friction characteristics of cast iron against cermet and asbestos-rubber materials, as well as steel and bronze in the alternating-sign friction are expressed by elliptical curves that can be satisfactorily described by a theoretical model [75, 82] and calculations [83]. The investigation results of work [84] prove that the friction force can be estimated by relation (5.22) only in the region of quasi-harmonic self-oscillations. In the region of relaxation self-oscillations the results show unacceptable errors, namely: the dynamic friction characteristic found by (5.23) is rising, whereas (5.22) gives it as falling.

Therefore, the experimental determination of function $f(v)$ for the non-uniform motion should be based on (5.23).

The analytical relation for the kinetic characteristic of friction allowing for the friction force dependence on acceleration has been first proposed by the authors of [85]

$$F_k = F_0 - k_1\dot{x} + k_2\dot{x}^2 + m_d \operatorname{sgn} \ddot{x}, \tag{5.27}$$

where $F_0$—Coulomb's friction, $k_1\dot{x}$—Newtonian viscous resistance, $k_2\dot{x}^2$—non-Newtonian viscous resistance inducing nonlinearity, $m_d$—actual part of complex mass:

$$m = m_d + im_u, \tag{5.28}$$

where $m_d$ and $m_u$—masses characterizing dissociation of energy and inertia in the system that are, correspondingly, the real and virtual parts of the complex mass. The relation of $m_u/m_d$ depends on the stress state, as well as crystalline and dislocational (dislocational type of density) structure of the surface layer.

Notice that actually the falling type of the kinetic characteristic is insufficient for the FS to be excited. This is supported by, e.g., the experiments performed by Kudinov and Lisitsyn [86], showing that self-oscillations might not be excited on the falling section of the curve but are present on the rising section of the characteristic.

### *5.3.2 Static Characteristics of Friction*

The static characteristic of friction is a dependence of the static friction force upon the stationary contact time. The first investigations devoted to this characteristic can be related to Coulomb's experiments. When studying the static friction coefficient for an oak sample rubbing against iron, this parameter was found to grow with time of the static contact (the static friction coefficient increases 2.3–2.4 times in 4 days). This growth was noticed to be more intensive when the normal load increased.

Coulomb's results were afterwards confirmed and elaborated for different materials of tribopairs in the works by Kragelskii [8], Kosterin and Kragelskii [34], Renkin [87], Hunter [88] and other researchers.

Such a behavior of the dependence is, apparently, because of the approach of the contacting surfaces, which increases the actual contact area and friction force, since the latter is the product of specific friction force $\tau$ multiplied by the actual contact area $A_r F = \tau A_r$.

We assume in the first approximation that τ remains constant when the surfaces are approaching. The second term in the binary dependence $\tau = \alpha + \beta q$ for soft materials constitutes a small portion, while coefficient $\beta$ values are about 0.01–0.02. Therefore, we accept the friction force growth to be proportional to the actual contact area increment in conditions of approaching surfaces in a dry friction mode. For the case of a plane rough surface interaction with a smooth one, the actual contact area is expressed as a function of the approach [8]:

$$A_r = A_c b \varepsilon^{\nu} \tag{5.29}$$

where $A_c$—contour contact area, $b$ and $\nu$—geometric parameters characterizing the shape of the supporting curve.

The contour contact area as well as constants b and v remain invariable, while relative approach ε equals during flattening to a relative deformation of the maximal by height asperities. So, when analyzing the variations in the friction force during formation of the contact, we first consider the deformation of separate asperities brought into contact in the overlap zone. The higher asperities undergo plastic deformation since even under a slight normal load the stresses acting on these asperities surpass much the yield point of the material being deformed due to a small actual contact area. This results in a mutual penetration of micro-asperities of contacting surfaces, which takes place even if two geometrically smooth surfaces are contacting because mechanical properties of the surface-forming elements are inhomogeneous. Because of the plastic flow of the materials in the contact zone the surface approach is a function of the normal load duration. Consequently, a relation is observed during friction between the friction force and rheological properties of contacting materials. Because of certain difficulties in adequate description of the mechanical behavior of solids, it seems impossible to construct an integral mathematical model to consider all peculiarities of plastic deformation of the materials even if we have at hand the required empirical data. In this connection, we have to employ some simplified models characterizing only the properties of the materials that are important for a given case.

One can describe the friction force variation in the course of contact formation by using the equations of linear viscoelasticity.

To analyze the creep and relaxation processes a number of researchers make use of Thompson's, Maxwell's and Ishlinskii's models.

The stress–strain relations for a 1D case are, correspondingly, the next:

$$\sigma = E\varepsilon + \eta\dot{\varepsilon}, \tag{5.30}$$

$$\frac{\sigma}{\theta} = \dot{\varepsilon} - \frac{1}{E}\dot{\sigma}, \tag{5.31}$$

$$r\sigma + \dot{\sigma} = Eu\varepsilon + \dot{\varepsilon}, \tag{5.32}$$

where $\sigma$—stress, $E$—elasticity modulus, $\varepsilon$—relative deformation, $\eta$—viscosity, $\theta$—relaxation time, $u$—speed of aftereffect, $r$—relaxation rate.

Since the processes at the initial stage of approach are similar to creep, we can employ (5.30–5.32). The analysis is, however, complicated because elementary asperities sequentially coming into contact are making stress on a single asperity fall down intensively. This is why the process of approach differs significantly from the creep that is characterized by a propagating with time deformation under a stable stress value.

The analysis of (5.30–5.32) presented in [8] shows that deformation of surface layers is satisfactorily described by Ishlinskii's (5.32) [89, 90]. We presumed to simplify the analysis that a single highest asperity participated in the approach. The effect of other smaller asperities coming gradually in contact was accounted for by stress variations on the highest asperity that occurred with the actual contact area growth, i.e.,

$$\sigma = \frac{N}{A_r}, \tag{5.33}$$

where $N$—normal load.

Solution of Thompson's (5.30), Maxwell's (5.31), and Ishlinskii's equations (5.32) in provision of conditions (5.33) gives us the corresponding relations

$$\varepsilon_t = \varepsilon_\infty - (\varepsilon_\infty - \varepsilon_0)\exp\left(-\frac{E}{\eta}\right), \tag{5.34}$$

$$\frac{N}{\theta A_a b}t = \frac{N_V}{A_a bE}\ln\frac{\varepsilon_t}{\varepsilon_0} + \frac{\varepsilon_t^{\nu+1} - \varepsilon_0^{\nu+1}}{\nu+1}, \tag{5.35}$$

$$t = \frac{\nu}{r}\ln\frac{\varepsilon_t}{\varepsilon_0} + \frac{\frac{r}{u}+\nu}{r(\nu+1)}\ln\frac{\frac{r}{u} - \frac{A_aEb}{N}\varepsilon_0^{\nu+1}}{\frac{r}{u} - \frac{A_aEb}{N}\varepsilon_t^{\nu+1}}, \tag{5.36}$$

where $\varepsilon_0$—asperity deformation (penetration depth) at zero contact time (the elastic component of approach); $\varepsilon_t$—strain arising as time $t$ expires.

Equation (5.35) admits qualitatively an infinitely increasing approach under a continuous loading, which contradicts the experimental data from [8, 91]. According to these data, the strain values and the static friction force tend to a certain limit. As it is seen from Fig. 5.6, relation (5.36) derived from Ishlinskii's equation describes most adequately the static friction force variations with the static contact time, which concerns also the approach of the surfaces. This is because Thompson's formula (5.35) is a particular case of Ishlinskii's relation at $\sigma = const$, $u = E/\eta$. In reality, the stress is quickly reducing during approach of the surfaces that makes plastic deformation less intensive, so the design data based on (5.36) are in better agreement with experimental evidences.

Formula (5.36) estimates the effect of separate factors on the approach and, consequently, on the actual contact area dependence upon the static contact duration. For instance, the normal load affects the approach dependence on time and, consequently, on the actual contact area growth. However, the increasing normal load, all other conditions being equal, leads to a more abrupt growth of the accrual contact area depending on the static contact time.

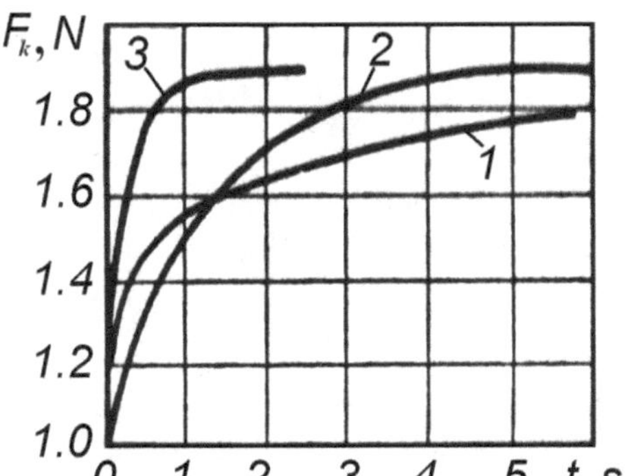


**Fig. 5.6** Comparison of experimental data (*1*) to calculations based on Ishlinskii's (*2*) and Thompson's (*3*) equations for a friction pair steel 45—polymethyl methacrylate (plexiglass)

It is evident from formula (5.36) that the dependence of the actual contact area versus contact time is strongly affected by geometrical dimensions of the contacting surfaces $A_a$, and surface roughness constants $b$ and $\nu$. This equation makes possible to trace the effect of physico-mechanical constants of the materials used in Ishlinskii relation upon the actual contact area.

If we neglect the first member of the equation, which effect is insignificant, we obtain:

$$\varepsilon_t = \left[\varepsilon_\infty^{\nu+1} - \left(\varepsilon_\infty^{\nu+1} - \varepsilon_0^{\nu+1}\right)\exp\left(-t\frac{r(\nu+1)}{\frac{r}{u}+\nu}\right)\right]^{\frac{1}{\nu+1}}. \tag{5.37}$$

For a frame model of the material subjected to surface hardening the dependence of relative approach versus loading at an infinitely long contact time is expressed by the following relation [8]:

$$\varepsilon_\infty = \left[\frac{N(\nu+\omega)}{A_c b H_y}\right]^{\frac{1}{\nu+\omega}}, \tag{5.38}$$

where $H_y$—constant characterizing plastic deformation; $\omega$—index of hardening.

Substitution of (5.37) in (5.29), gives us a formula describing the actual contact area dependence on the static contact time:

$$A_r = A_c b\left[\varepsilon_\infty^{\nu+1} - \left(\varepsilon_\infty^{\nu+1} - \varepsilon_0^{\nu+1}\right)e^{-t\frac{r(\nu+1)}{\frac{r}{u}+\nu}}\right]^{\frac{1}{\nu+1}}. \tag{5.39}$$

It is evident from (5.39) that the actual contact area dependence on time is conditioned by the following factors: relaxation rate $r$, aftereffect $u$, and geometrical constants $b$ and $\nu$. The analysis of (5.39) shows that the contact area is initially increasing intensively, then the growth is retarded and afterwards the contact area tends to some constant value.

We may neglect the second member in the binomial law of friction, whereupon the specific friction force $\tau$ will become constant. So, we can find the friction force variation with contact time duration by substitution of (5.39) in (5.13):

$$F = \tau A_c b\left[\varepsilon_\infty^{\nu+1} - \left(\varepsilon_\infty^{\nu+1} - \varepsilon_0^{\nu+1}\right)e^{-t\frac{r(\nu+1)}{\frac{r}{u}+\nu}}\right]^{\frac{1}{\nu+1}}. \tag{5.40}$$

This equation is similar in structure to the one obtained in [92] in supposition that $\sigma = const$ and the friction force is directly proportional to deformation

$$F_s(t) = F_\infty - (F_\infty - F_0)e^{-ut}, \tag{5.41}$$

where $F_s(t)$—static friction force upon expiration of time $t$ of the static contact, $F_\infty$—friction force at infinitely long contact time, $F_0$—friction force at zero contact time, $u$—constant characterizing the strengthening rate of the bond.

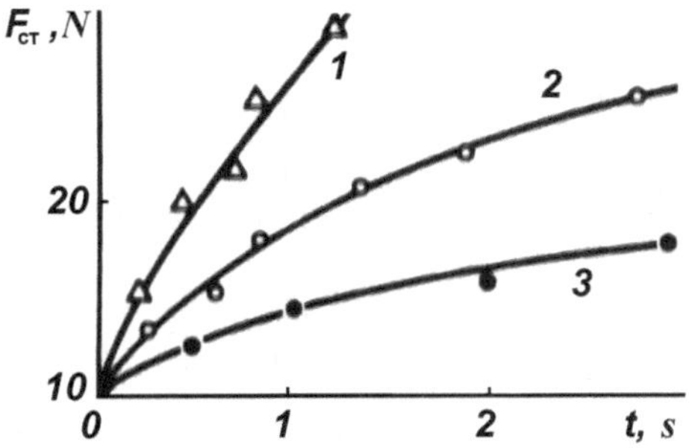


**Fig. 5.7** Static friction force dependence on static contact duration under different normal loads for the friction pair steel—plexiglass: *1*—100 N; *2*—50 N; *3*—30 N

The investigation results presented in [89, 90] make it possible to estimate the effect of such factors as the normal load, geometrical parameters of surfaces, physico-mechanical properties of the rubbing materials upon the static characteristic of friction. Figure 5.7 illustrates the correlation between the design and experimental evidences characterizing the static friction force variation with time for different normal load values. It is also shown that geometrical constants $b$ and $v$ effect differently the friction force variation. The friction force increases intensively as constant $b$ grows, and tends to diminish as constants $A$ and $v$ increase, all other conditions being equal.

The effect of physico-mechanical properties of materials is also different. The materials with a high elasticity modulus, and aftereffect speed but a low relaxation rate allows for a less intensive growth of the friction force with time.

The analogous characteristics of the friction force were obtained by Kosterin and Kragelskii [34] for the friction materials on the polymer matrix base. Figure 5.8 shows typical static characteristics of some friction materials used in brakes and couplings [8]. The investigations have proved that the static friction coefficient growth with time of a stationary contact in conditions of elevated temperatures can be attributed not only to strengthening of the formed bonds but also to their increment in number. This growth is less intensive in the materials based on the resin binders in contrast to the rubber-based ones.

The formula expressing the static friction force dependence on the stationary contact time differs a little from (5.41) [34]:

$$F_s(t) = F_\infty[1 - \exp(-k_1 t^u)]^{k_2}, \qquad (5.42)$$

where $k_1$ and $k_2$—empirical coefficients.

Formulas (5.41) and (5.42), comply well at a rather large $t$ but make no account for the normal contact stress variations in time. The stresses were taken in

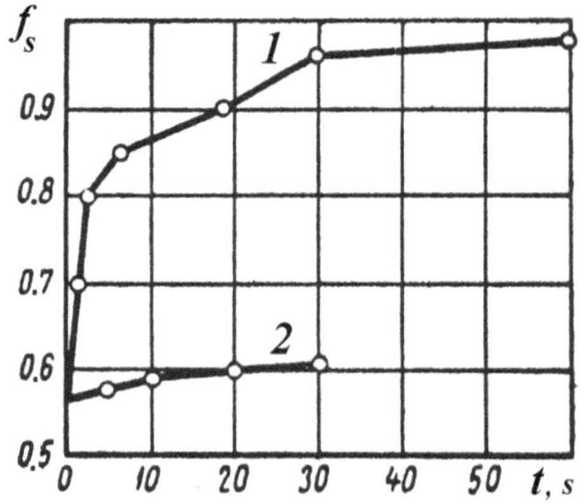


**Fig. 5.8** Static friction coefficient versus time of stationary contact for different frictional materials: *1*—rubber binder-based frictional material; *2*—resin-based frictional material

consideration by the authors of [93] for the case of predominantly plastic deformations in the contact. The friction force was found by a static summation of elementary friction forces on separate rough surface regions:

$$F_s(t) = k_c \frac{\tau}{p} N, \tag{5.43}$$

where $k_c = A_c/A_e$, $A_e$, $A_c$—actual elementary contact and shear areas; $\tau$ and $p$—tangential and normal stresses on the elementary contact; $N$—force perceived by all elementary contacts.

Time dependence of $\tau$ and $p$ is shown as a sum of constants $\tau_0$, $p_0$ and varying parts:

$$\tau = \tau_0 + \frac{1}{\ln k_\tau} \ln\left[1 + \frac{k_{\max}(k_\tau - 1)y}{k_{n\pi} k_{\max}(1 - y)}\right] \tag{5.44}$$

$$p = p_0 + \frac{1}{\ln k_p} \ln\left(1 - ke^{-\delta t}\right) \tag{5.45}$$

where $k_\tau$—deformation factor; $k_{\max}$—maximum irreversible shear; $k_{n\pi}$—plasticity coefficient; $k_p < 1$, $k < 1$, $\delta < 1$—plasticity constants $y$—time-dependent relative compression.

Substitution of (5.44) and (5.45) in (5.43), gives a relation for expressing the friction force as a function of time, which is positive at whatever parameter we take

$$F_s(t) = k_c \frac{\ln k_p}{\ln k_\tau} \frac{\tau_0 \ln k_\tau + \ln\left[1 + \frac{k_{\max}(k_\tau - 1)y}{k_{n\pi} k_{\max}(1 - y)}\right]}{p_0 \ln k_p + \ln(1 - ke^{-\delta t})} N \tag{5.46}$$

This function presents special interest at $t = 0$ and $t = \infty$:

$$F_0 = k_c \frac{\tau_0}{p_0} N; \tag{5.47}$$

$$F_\infty = F_0 + k_c \frac{\ln k_\tau + \ln\left[1 + \frac{k_{\max}(k_\tau - 1)}{k_{n\pi}(1 - k_{\max})}\right]}{p_0 \ln k_\tau} N. \tag{5.48}$$

It follows from (5.48) that the coefficient of stationary friction is determined by the plastic properties of the materials and is independent of surface microgeometry.

The theoretical calculations of the static friction characteristic that accounts for the electrostatic forces in the contact between glass surfaces have brought us to a relation similar to (5.41) [94].

The static characteristic of friction is presented in a number of works as a power function of time. A formula is proposed in [95] proceeding from a supposition that the force able to shear micro-asperities on rough surfaces augments with increasing tangential microshear

$$f_s = f_k + k_1 t^{k_2} \tag{5.49}$$

where $0 < k_1$, $k_2 < 1$—coefficient of power approximation.

The increase in the static friction characteristic is attributed to the normal contact deformation events expressed in the material creep in the case of plastic deformations [136]. The equation for the deformation process expressed through a thermally activated self-diffusion is

$$f_s = f_k + k_3 \exp\left[-\frac{E_a}{RT}\left(\frac{v}{k_4 v + 1}\right)\right] t^{\frac{v}{k_4 v + 1}} \tag{5.50}$$

where $E_a$—self-diffusion activation energy, $R$—universal gas constant, $T$—absolute temperature, $k_4$—constant, $v$—parameter of power approximation of the initial piece of the bearing profile curve [96]. In isometric conditions (5.50) is simplified till (5.49) where the coefficients are dependent on contact conditions and elasto-plastic properties of the friction pair materials.

If we express the static friction characteristic in terms of the power function, the friction force will infinitely grow, thus contradicting the reality.

This kind of a stick-slip transfer from the static to sliding friction is explained in [97] by the microshear ordering as well as by the effect of hardening and rest processes on the rubbing material surfaces. The conjugated materials at static friction are subjected with time to plastic deformation in each elementary microcontact. The shear occurs in all microcontacts in the moment of transfer to sliding under the maximal elementary shear force value. Further sliding is characterized by formation of new microcontacts and shear of the previously formed ones, i.e., ordering of the shear events is violated and they occur on the microcontacts not simultaneously. Besides, when transferring to sliding the material undergoes hardening as a result of increasing deformation work that reduces the friction force. So the relation of the static to kinetic force can be written as follows [97]:

$$F_s = k_{y1} k_{y2} F_k, \tag{5.51}$$

where $k_{y1}$—coefficient of material strengthening, $k_{y2}$—ordering factor dependent on shear rate.

The friction force versus contact time dependence is presented in [96] by a similar (5.13):

$$F = \tau\, t_m A_c [\varepsilon(t)]^{\nu}, \tag{5.52}$$

where $t_m$—bearing profile length on the median line level, $A_c$—contour contact area, $\varepsilon(t)$—viscoelastic deformation in the contact.

The dependence of viscoelastic deformation on time is obtained from the relation

$$\varepsilon(t) = \varepsilon_y + \varepsilon_n \left[1 - \exp\left(-\frac{t}{t_p}\right)\right], \tag{5.53}$$

where $\varepsilon_y$ and $\varepsilon_n$—elastic and plastic components of deformation, correspondingly, $t_p$—time of deformation relaxation under a constant stress.

By substitution of (5.53) in (5.52), we come to the friction force versus contact time relation, which is common for the static and kinetic friction modes:

$$F = \tau t_m A_c \left\{ \varepsilon_y + \varepsilon_n \left[1 - \exp\left(-\frac{t}{t_p}\right)\right]^{\nu} \right\}. \tag{5.54}$$

In the case of a prolonged immovable contact duration and infinitesimal life of the elementary actual contact sites at sliding, the static friction force relation to the sliding one is found from the relation of the total elastic plus plastic deformations to the elastic one:

$$\frac{F_\infty}{F_k} = \frac{\varepsilon_y + \varepsilon_n}{\varepsilon_y}. \tag{5.55}$$

## 5.4 Self-oscillation Mechanism in Metal–Polymer Friction Pairs

### 5.4.1 Adhesive Mechanism

High contact pressures arising on the actual contact spots may lead to formation of the local adhesive bridges of welding. A number of researchers attribute the stick-slip friction behavior of the microseizure–microslip type to formation and breakage of these bridges [39]. The conjugated surface areas of the rubbing bodies are moving conjointly for some time during microseizure until the growing external force becomes sufficient for a shear followed by sliding and further accelerated microslip till formation of the next local bridge of welding. The mechanisms of static friction and sliding are assumed roughly similar. They differ mainly in unequal efficient time of contact. The rising FS level with the speed growth is explained by the increasing life of the adhesive welding bridges.

The FS are known to depend on tribological properties of the contact, which, in their turn, are a function of the external factors like the sliding velocity, eigenfrequency, design of friction units and damping.

The friction force and FS can be reduced by using a lubricant and the degree of this reduction is conditioned by antifrictional properties and oxidation level of the LM. As it follows from (5.12), the usage of the LM allowing for a friction coefficient parity in statics and sliding, makes the FS improbable. In the case the boundary lubrication mode prevails within the low-velocity region, there appears a possibility of the oil film formation between the sliding surfaces. This leads to the appearance of a partially hydrodynamic lubrication on individual contact spots, where the friction force falls. So, the friction force is rather quickly reducing as the velocity increases, which makes motion nonuniform. The formation of quasi-hydrodynamic lubrication is affected by the lubricant film viscosity, contact area, friction surface microgeometry, shear rate gradient, and some other factors, altogether hampering the quantitative estimate of the results.

The influence of tribological properties of structural and lubricating materials has been studied at length experimentally in [98]. Molecular interactions of the rubbing surfaces as the cause of leaps in conditions of starvation lubrication and dry friction have been examined in [99].

According to [3, 100], thermal regime in the contact zone is the major factor that defines the type of the static and kinetic characteristics and, consequently, the conditions of the FS generation. The authors of [101] propose an explanation to the FS proceeding from the attenuating dependence of the friction force versus temperature. They believe that the friction surface heating during the leap reduces the friction force and leads to an elastic unloading. As a result, the surface cools down during deceleration (stoppage). Consideration of the mean surface temperature in this model enables to calculate and substantiate the FS only in the severe friction conditions, while consideration of the temperature flash on the actual contact spots [102] helps to study the FS in the common frictional systems under the low sliding speeds.

### *5.4.2 Synchronization of Frictional Micro-Oscillators*

It was shown in [103] that for metal-polymer pairs rubbing without lubrication the abrasive wear mode dominates on the initial stages of friction (run-in stage). Debris particles formed at this stage of friction are mainly the result of the polymeric

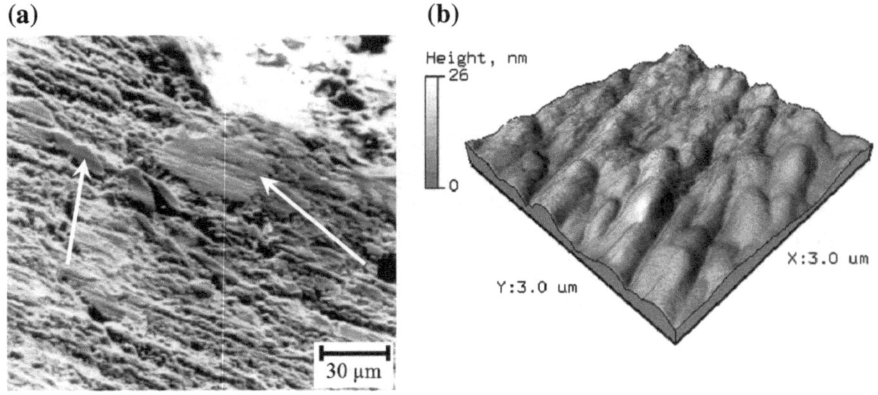


**Fig. 5.9** Contact surfaces of metal-polymer friction pair: SEM images of laminated solid structures formed on frictional material surface (**a**); AFM images of friction surface of metal counterbody (**b**)

matrix wear. A part of the debris removed from the friction zone are transferred into the environment, the other part remains on the counterbody friction surface. The remaining particles attach to the friction surface by way of impressing into a less hard matrix of the friction composite. This is the reason, why the wear rate reduces essentially some time later [103–105]. As a consequence, the normal and tangential stresses start to increase with the temperature in the actual contact zone. In majority of modern brake systems such local stresses and loads might reach the magnitude under which surface layers of the friction composites undergo melting with formation of a fine layered structure. Physico-mechanical properties of the latter (hardness, elasticity modulus) differ much from those of the volume properties of the initial matrix [106]. SEM images of the friction surface of the polymeric material with the formed laminated surface structures after rubbing against a metal counterbody are presented in Fig. 5.9a [103]. A specific morphology of the metal counterbody contact surface characteristic for the appearance of the high-frequency FS (10–16 kHz) is shown in Fig. 5.9b.

So, we have isolated a few types of contact zones and corresponding scenarios of contact interactions on the friction surface presented schematically in Fig. 5.10 [107]. The first type of the contact zones is characterized according to Fig. 5.10 by interactions of abrasive particles, the second type by interactions of solid lubricants (friction modifiers), the third one of the solid surface layers (third-body), and the fourth type by interactions between the metallic fillers. However, the main energy portion scattered during braking is spent on the third type of the contact zone (adhesive type) [103]. The regions found in the vicinity of the laminated structures with elevated hardness are subjected to a more intensive wearing, which together with intensified heat generation and high mechanical stresses leads to nucleation and propagation of cracks in the subsurface microlayers.

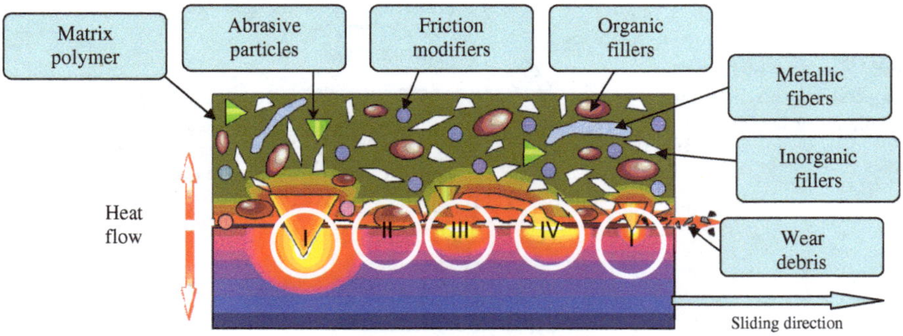


**Fig. 5.10** A tribological scheme of "frictional composite—metal counterbody" interaction [107]

Contact interactions between microasperities on the metal counterbody surface during relative tangential displacement of the rubbing bodies are exciting tangential oscillations of the third body layers and intensify the processes of subsurface cracking, leading to their gradual delamination and formation of debris particles. The formation and failure kinetics of these microlayers has been studied in detail in what concerns the surface structure generation [108, 109] and wear [110] using the method of the finite-state machine (FSM), popular today in studies of tribosystems. In particular, it was proved that the described solid microlayers present a stable self-organized mesolevel system [108–110].

The dynamic equilibrium occurred in formation and failure processes of the solid surface structures turns to be the main factor that governs the wideband dynamic loading of the friction contact. It also serves as an energy source for the FS arising at frictional interactions in metal-polymer pairs.

Bowden and Tabor [98] were the first to discover that surface microasperities are preserving their tendency to the elastic microdeformation even in an ideally rigid frictional system. They have also come to a conclusion that elasticity of asperity tips can be a source of self-oscillations.

Modern representations on excitement of FS in metal-polymer friction pairs proceed from the consideration of interrelated elementary oscillators (separate sources of friction-induced micro-oscillations and acoustic emission pulses).

These oscillators are forming during contact interactions between solid surface structures of the frictional material and a rough metal counterbody surface, schematically presented in Fig. 5.11. Since a lion share of the friction energy in the given system is generated through the adhesive type of interactions [111], we shall further analyze micro-oscillations of this type of interactions only.

In our case, we may neglect the variations in size and number of the contact zones since the processes of failure and regeneration of the solid surface layers are taking much longer time as compared to the vibroactivity time scale of the elementary oscillators under study. If we take into account above-cited mechanism of contact interactions between solid surface structures of the frictional material and metal counterbody asperities, the dynamic scheme of a vibration-active friction pair

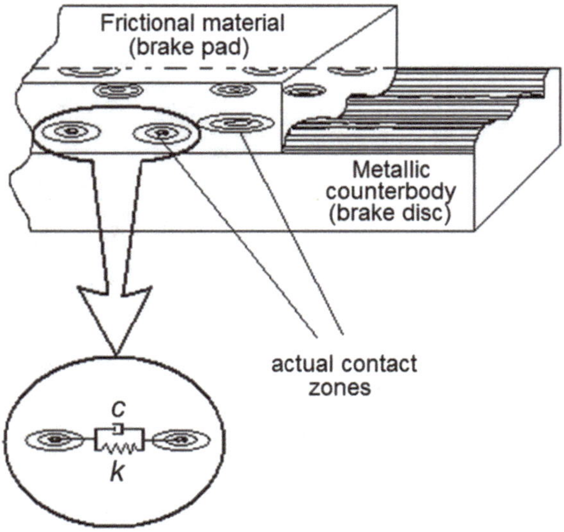


**Fig. 5.11** Contact interaction scheme with formation of single sources of friction-induced microoscillations (elementary oscillators)

can be presented in the form of the contacting layers and microvolumes, i.e., the frictional elements (oscillators with distributed mass, stiffness and viscosity parameters) experiencing the corresponding normal ($N_I$, $N_{II}$, …, $N_V$) and tangential ($F_I$, $F_{II}$, …, $F_V$) forces induced by the external and internal friction relative to a single oscillator. Thanks to the polymeric matrix elasticity, each element is elastically linked with a solid surface layer in both normal and tangential directions, all elements being also interconnected as is shown in Fig. 5.12.

We suppose that the mechanism of noise and vibration generation at frictional interaction between solids within the frames of above-proposed model is the next. Multiple elastic and inelastic impulse redeformation events of microasperities, surface areas and microvolumes in the surface layers along with competing

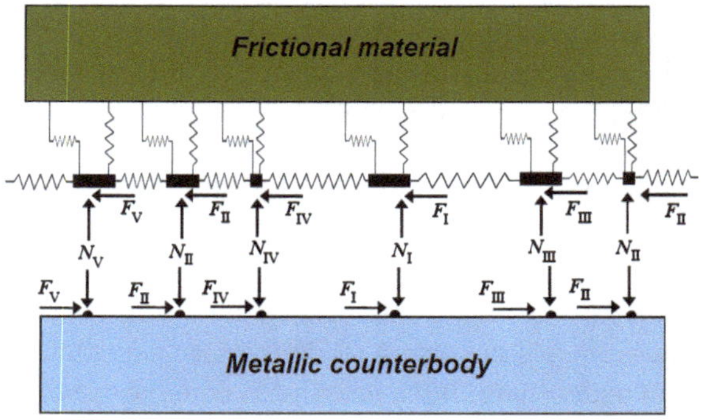


**Fig. 5.12** Dynamic scheme of a vibroactive friction pair presented as a combination of interrelated elementary oscillators on actual contact spots [112]

processes of wear and regeneration of the fine films generate wide-range micro-oscillations of the contacting surfaces. The elastic links between elementary oscillators characterized by dynamic properties of the frictional composite (dynamic elasticity modulus and loss factor) make probable synchronization of the elementary oscillators generating micro-oscillations at the frequencies commensurable to those of a mechanical friction system [111].

The probability of synchronizing self-oscillations is referred to as a fundamental property of nonlinear systems. The problems of mutual synchronization are treated thoroughly in the corresponding chapters of the theory of nonlinear dynamics [113–115]. It should be noted that mutual synchronization of micro-oscillators in the system under study increases the amplitude of synchronous micro-oscillations. As a result of mutual effect of dynamic processes in the friction zone and in the whole mechanical system, there arises frictional instability at the macrolevel that is usually accompanied by the high-level vibrations and/or intensive noise of the friction joint [115, 116].

### *5.4.3 Interrelation of Normal and Tangential Micro-Oscillations*

The spectral and time analyses of instantaneous friction force values on the surface [117] have shown a random dynamic loading of the contact and frictional instability caused by self-oscillations of interacting microasperities. Their contact deformation induces normal and tangential oscillations that are dependent on the friction conditions, surface wavelength spectrum and its variation behavior. Oscillations of the normal and friction forces either coincide or are close in phase, and their spectra are in fact similar. The mean quadratic value of the normal force fluctuations $F_N$ is weakly dependent on the mean normal load. It increases with increasing sliding velocity $\vartheta$ and reducing roughness $R_a$:

$$F_N = k\frac{\vartheta^{1,5}}{R_a}, \tag{5.56}$$

where $k$—coefficient dependent on material properties and friction conditions.

Introduction of the LM into the friction zone promotes damping of self-oscillations, especially the HF ones due to the surface roughness smoothening first of all. In these conditions, the dynamic loading of the contact remains the same, although the oscillations of the normal force are by an order of magnitude less in the presence of at least minimal boundary lubrication than without it [117].

The elementary oscillators receive micropulses from microasperities of the sliding counterbody. The normal components of these micropulses are continuously exciting vibrations of the rubbing bodies in the normal direction. The main frequency of these self-oscillations in dry friction are about a few thousand Hertz, which are dependent, first of all, on contact stiffness. Low amplitudes and insignificant resonant bandwidth of the normal self-oscillations are the reason of their late detection [118, 119].

The role of normal displacements in the mechanism of FS excitement was predicted as far back as in the 1940s and 1950s of the 20th century. However, the first theories for the FS with account of both tangential and normal self-vibrations were developed by Kudinov [16] for the semifluid lubricants and by Tolstoy and Kaplan [120] for dry friction.

Named theories are considering the mechanism of FS generation as follows. Random variations in sliding conditions (velocity, acceleration, etc.) induce deformations of the drive and promote the tangential shear. The latter brings about a normal deformation relative to the friction surface that alters the friction force, and a tangential deformation of the elastic elements of the system. Certain phase ratios of the tangential to normal shear (oscillations) create the conditions when the friction force varies synchronously with longitudinal tangential vibrations.

Generation of the FS at dry friction or boundary lubrication can be simulated in provision that cross dimensions of the rubbing element do not cede much its height. In this case, the elastic compliance in the normal direction is by several orders less than tangentially. Proceeding from this fact, the authors of [73] have proposed the following model. An element of a friction pair was assumed to be an ideally rigid body resting on a system of springs that simulate microasperities (see Fig. 5.13). The given scheme differs from the known rod model [8] in the use of springs instead of the rigid rods. The long springs bear the normal load, the shorter ones do not participate in the interactions. Any normal pulse may generate free oscillations of the weight in the normal direction. These oscillations are nonlinear and asymmetric since more and more springs start to accept the load as the weight descends. Therefore, stiffness of the contact is unstable: it grows as the slider descends and diminishes with its ascending. Growth of the amplitude of oscillations due to their asymmetry elevates the mean level of the weight over the counterbody, reduces the mean number of asperities contacting at a time and their total area (actual contact area), altogether decreasing the friction force.

It can be traced that the higher is the sliding velocity, the more intensive are the normal components of micropulses between microasperities and the larger is the

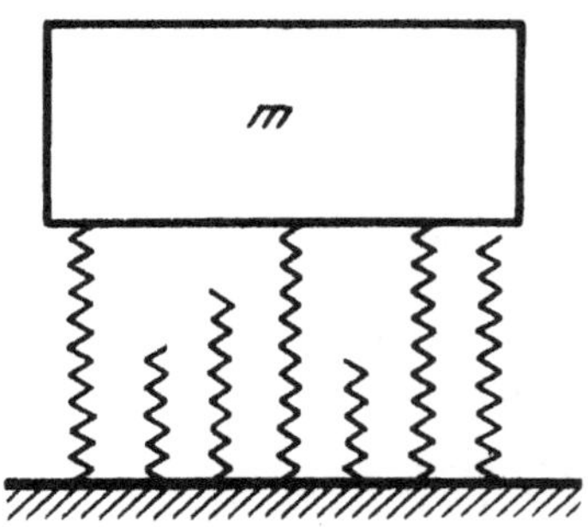


**Fig. 5.13** Model of contact interactions on springs [73]

oscillation amplitude (almost linear dependence) and, correspondingly, the smaller is the friction force. The latter is a factor explaining the falling kinetic characteristic in dry friction or boundary lubrication.

The representation that oscillations are coupling in frictional systems, meaning that the normal and tangential, longitudinal and transverse self-oscillations are interrelated, has been introduced first by Kudinov [121]. This coupling depends upon the vicinity of the main frequencies in the normal $\omega_{fn}$ and tangential $\omega_{f\tau}$ directions estimated approximately by eigenfrequencies of linear oscillations [122]:

$$\omega_{fn} = \sqrt{\frac{k_n}{m}}, \quad \omega_{f\tau} = \sqrt{\frac{k_\tau}{m}}, \tag{5.57}$$

where $k_n$ and $k_\tau$—normal and tangential contact stiffness values, correspondingly.

Because of intricacy and stochastic nature of the actual friction processes, the spectra of acoustic oscillations of contacting surfaces turn to be polyharmonic. A distinguishing feature of such oscillations is the presence of multiple simultaneously excited acoustic sources of different power that are randomly scattered over the nominal contact area. Processing of their signals using the harmonic analysis has shown the existence of the main frequency found from formula (5.57) with an error ≤10 % [122]. Nevertheless, the relationship between the parameters of acoustic emission and frictional interactions may be described reliably only if we take into account all the sources, which is actually improbable. In some cases this can be reached only through sequential approximations and idealization procedures [123].

### *5.4.4 Analysis of Oscillations of an Elementary Unbound Oscillator*

When developing the facilities intended to hamper vibroacoustic activity from its very beginning, one should analyze in detail the tangential friction-induced micro-oscillations in order to find the structural factors able to abate instability at the microlevel.

In this connection, we take an idealized dynamic scheme of a vibroactive friction pair and analyze it at the phenomenological level. The phenomenological model of an elementary unbound oscillator is illustrated schematically in Fig. 5.14. An elementary oscillator is presented as an oscillating system consisting of mass $m$, elastic element $k$ and the element of internal friction losses $c$. Named units are simulating the inertial properties of a solid surface layer of a frictional composite bound in tangential direction with a viscoelastic polymeric matrix of the frictional composite of a rigidly fixed lining. These layers on the real contact surface of the frictional composite are shown by arrows in Figs. 5.2a and 5.9a. The elementary oscillator is affected in a single contact interaction event by the normal force $N$ and friction force $F$ directed opposite to the metal counterbody motion. The oscillations are excited

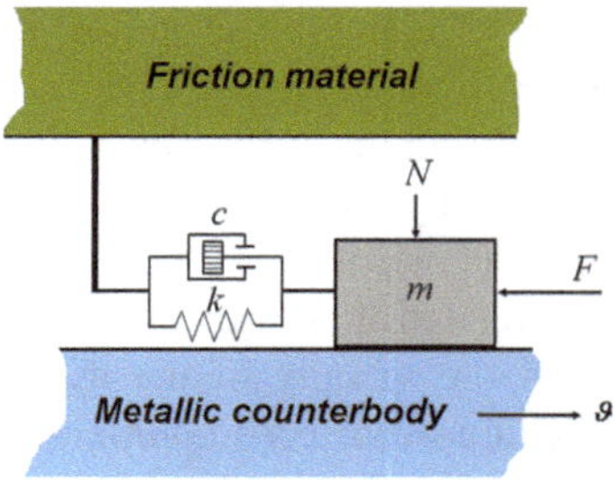


**Fig. 5.14** Phenomenological model of elementary oscillator [110]

by a series of contact interactions between solid friction surfaces of the FM and microasperities of the metal counterbody moving at velocity $\vartheta$ relative to a stationary lining.

The equation for the motion of an elementary unbound oscillator with account of above model is of the form:

$$m\frac{d^2x}{dt^2} + kx = F - c\frac{dx}{dt}, \tag{5.58}$$

where $x$—coordinate, $F = \mu(\vartheta_r)N$—friction force, $\vartheta_r = \vartheta - \frac{dx}{dt}$—sliding velocity of an elementary oscillator relative to the original position (unbiased).

To express the friction coefficient dependence on sliding velocity we can use the following relation [111]:

$$\mu(\vartheta_r) = \frac{0.4}{\pi}\operatorname{arctg}(200 \cdot \vartheta_r)\left(\frac{1}{|\vartheta| + 1} + 1\right). \tag{5.59}$$

It should be noted that above-mentioned linear system is related to self-vibrating ones. We have paid special attention in this work to (5.58) for a self-oscillating system that accounts for the forces of internal damping $F_{in}$, as a function of viscoelastic properties of the polymeric matrix

$$F_{in} = kx + c\frac{dx}{dt}. \tag{5.60}$$

Based on the known mathematical procedures for self-oscillating systems [13, 124], it is possible to prove that motion of a system may vary significantly in response to the forces of internal damping (as opposed to the external friction force), ranging from the uniform till a stick-slip motion like that in Fig. 5.15.

Solution of (5.58) that considers both static and dynamic characteristics of external friction for the case of relaxation oscillations where the static friction exceeds the sliding one, has shown that there exists some critical velocity of the counterbody $\vartheta_k$ below which the motion is exclusively of a jerky character, and above which, the self-oscillating mode of friction is improbable [1, 23]

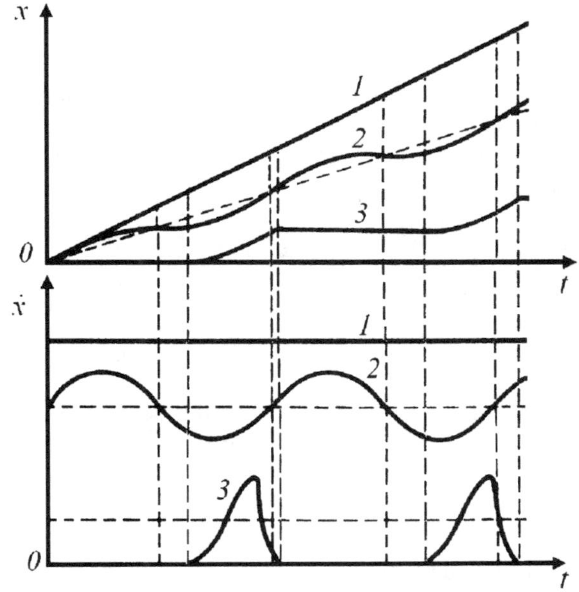


**Fig. 5.15** Types of sliding: *1*—uniform sliding ($F_{вн} \gg F$); *2*—harmonic oscillations ($F_{вн} \approx F$); *3*—relaxation oscillations ($F_{вн} \ll F$)

$$\vartheta_k = \frac{\Delta F}{\varphi_c \sqrt{mk}}, \tag{5.61}$$

where $\Delta F$—difference between the static friction forces and sliding ones defined by the static and kinetic characteristics of friction, $\varphi_c$—function interrelated with the vibration damping factor in an inexplicit form

$$\theta = \frac{c - \alpha_k}{2\sqrt{m/k}}, \tag{5.62}$$

where $\alpha_k$—slope of frictional curve falling shown in Fig. 5.1 at small θ values

$$\varphi_c \approx \sqrt{4\pi\theta} \tag{5.63}$$

However, the procedure where the vibration activity of a friction pair is reduced by reducing the friction force or by varying statico-kinetic characteristics via adjusting the friction coefficient or the load/velocity parameters is often admissible in respect of attaining a needed frictional efficiency or even improbable in conditions of nonstationary friction.

The analysis of the equations of motion for an elementary oscillator and understanding of the mechanism of noise and vibrations, has brought us to the assumption that one of the major factors (criteria) affecting the probability of instability onset (transfer into the macro-oscillation mode) on the friction contact is a tribosystem ability to damp (attenuate) friction-induced tangential micro-oscillations.

### 5.4.5 Contact Damping

To achieve quantitative relations in order to estimate damping capacity of the FM in terms of the proposed phenomenological model, let us consider the equation describing vibrations of an elementary oscillator (5.58) for the case of harmonic oscillations

$$F_{вн} \approx F.$$

If we substitute the exciting force in an exponential form

$$F = F_0 e^{i\omega t},$$

and use the equation for sinusoidal processes in a complex form, the equation for vibrations of an elementary oscillator will be as follows:

$$x = x_0 e^{i\omega t}, \tag{5.64}$$

where $x_0$—amplitude of vibrations, ω—circular frequency, $t$—time, $i = \sqrt{-1}$.

After introduction of the value

$$\eta = \omega \frac{c}{k},$$

Equation (5.58) can be written in a complex form

$$kx + c\frac{dx}{dt} = (k + i\omega c)x = k(1 + i\eta)x. \tag{5.65}$$

The losses of mechanical energy characterizing damping capacity of the frictional contact can be considered by introduction of a complex stiffness $k(1 + i\eta)$. Since stiffness is proportional to the elasticity modulus of the material, we can accept that

$$E = E'(1 + i\eta) = E' + iE'', \tag{5.66}$$

where $E'$ and $E''$—actual and imaginary parts of the complex elasticity modulus.

The real part $E'$, called the dynamic modulus of elasticity $E_d$, shows stress to strain relation in the material during its vibration. $E_d$ is a dynamic analogue of the static modulus of elasticity $E_s$.

Commonly, the dynamic modulus of elasticity for the composite materials on the polymer matrix base is several times higher than the static one. The difference between $E_d$ and $E_s$ grows with increasing porosity of the composite. Besides, the dynamic modulus of elasticity of the composites with the porous, fibrous, or fibrous-porous structure depends much upon loading on the material and frequency [125, 126]. The difference between $E_d$ and $E_s$ is explained as follows. Under the influence of a periodic force the deformation lags behind the force due to intrinsic for the viscoelastic materials hysteresis. Therefore, in the moment corresponding to a maximal force the dynamic deformation will be less than the static one by some positive number $a$. So far, the dynamic stiffness $k_d$ will exceed the static one

$$k_d = \frac{F}{x_s - a}. \tag{5.67}$$

The degree of dynamic stiffness prevalence over the static one and, correspondingly, $E_d$ over $E_s$ depends on the internal dissipative properties of the material being a structurally sensitive parameter.

The imaginary part in (5.66) characterizes irreversible losses (dissipation) of the mechanical energy as a result of internal friction during vibration. The ratio $\frac{E''}{E'}$ equals to a tangent of the shear angle between the stress and relative strain. It is the larger the higher are the energy losses on the internal friction. The parameter that characterizes energy losses on the internal friction is either a tangent of the angle of mechanical losses $tg\delta$ or a loss factor $\eta$.

Proceeding from above-considered mechanisms and experimental data on the effect of dynamic mechanical characteristics of frictional composites on vibroacoustic activity of tribojoints, it is proposed to use a relative parameter $D$ found from relation (5.68) [127] as a factor characterizing damping capacity of FM. It is intended to reduce tangential micro-oscillations of the surface areas and microvolumes in superficial layers (elementary oscillators in a general case) and prevent transfer of the system into a frictional macroinstability as a result of mutual synchronization

$$D = E_d \eta, \tag{5.68}$$

where $E_d$—dynamic elasticity modulus of FM, $\eta$—loss factor of the material.

The proposed parameter of damping capacity with consideration of given friction characteristics of a tribopair assists in making express-estimates and prediction of its vibroacoustic activity. Besides, it is used in the present work as an optimization criterion of the dynamic mechanical characteristics of the FM at the stage of designing [127].

## 5.5 Calculation of Friction-Excited Self-oscillations in Macrosystems

Figure 5.16 illustrates a scheme of a frictional system analogous to the one described by Block in [44]. It includes weight 2 of mass $m$ connected to an elastic element (spring) 3 of spring rate $k$ on a moving with velocity $v$ belt 1. Stiffness of the belt is much higher than that of the elastic element. Damping $c$ is accounted for by the element 4.

Motion of the weight in the given system is described by a heterogeneous equation

$$m\ddot{x} + c\dot{x} + kx = F \tag{5.69}$$

However, solution of this equation presents certain difficulties, wherefore we propose a graphical solution in the form of a relationship between two dimensionless parameters of the system. It follows that stability of the system can be reached through varying mass of the moving body, rigidity and damping parameters of the system. Thus, we have derived the dependencies interrelating the vibratory load with other parameters of the system and the latter in between

$$q = \frac{\vartheta\sqrt{km}}{F_{cm}} \tag{5.70}$$

and damping

$$D = \frac{\beta}{2\sqrt{km}} \tag{5.71}$$

However, H. Block did not make allowance for the kinetic characteristic of friction, reckoning that the amplitude of oscillations equals to a double difference of the static and dynamic friction coefficients. Besides, he attributed the amplitude decay as the velocity increases to damping in the system.

The theory of friction-induced self-oscillations that accounts for the static characteristic of friction was put forward first by Ishlinskii and Kragelskii in work [90]. They have shown that the static friction force in a frictional system is decreasing with the stationary contact duration like in the one presented in Fig. 5.16. What is more, the type of the static characteristic of friction depends most of all on the rheological properties of the friction pair materials. This is an explanation to the fact that the amplitude of the first oscillation at transition to sliding is larger than the succeeding ones (Fig. 5.17). I.V. Kragelskii and Yu.I Kosterin have further elaborated this theory [34] with account of the kinetic characteristic for the analogous frictional system, although without damping consideration.

The weight in the system under study is found in the state of rest for a rather long time, after which the belt starts moving at a constant velocity $\vartheta$, stretching thereby

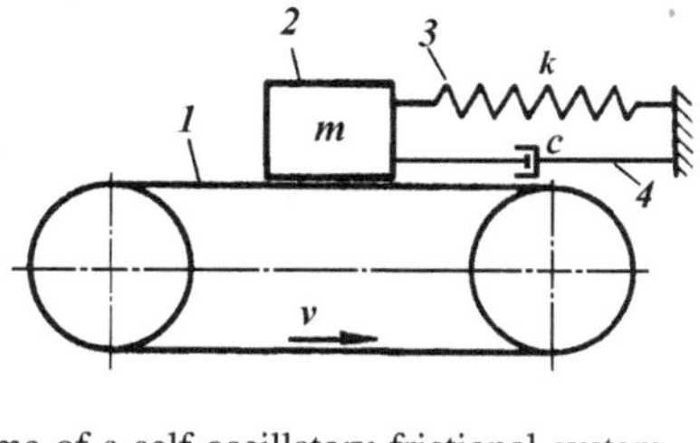


**Fig. 5.16** Equivalent scheme of a self-oscillatory frictional system

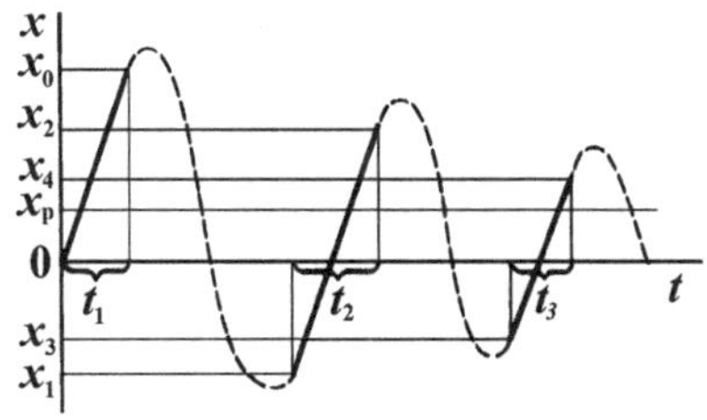


**Fig. 5.17** Scheme of oscillating motion of weight $m$ (see Fig. 5.16)

the spring. In the moment its tension force becomes equal to the maximal force of static friction $F_{\infty}$, the weight starts moving relative to the belt:

$$kx_0 = F_{\infty}, \tag{5.72}$$

where $x_0$—shift at initial sliding of the weight. Further on, the weight displaces under the elastic force of the spring and sliding force: $F_k < F_{\infty}$, $F_k = \text{const}$. This motion is oscillatory within the vicinity of equilibrium $x_p$ and is determined by the relation

$$kx_p = F_k. \tag{5.73}$$

Initially, the weight moves at velocity $\vartheta$. It displaces over the surface till its velocity becomes equal to $\vartheta$ in coordinate $x_1$ again, which s illustrated in Fig. 5.17:

$$x_1 = 2x_p - x_0. \tag{5.74}$$

Further sliding of the weight becomes impossible in point $x_1$ since the force changes its sign, and its value turns to be larger the elastic force of the spring. The sample initiates motion again together with the plane till moment $t_2$ at which the spring force becomes equal again to the static friction force in point $x_2$:

$$k(x_1 + vt_2) = F_{cm}(t_2), \tag{5.75}$$

where the value of $F_{st}$ till the moment of shear of the sample will depend on the static contact duration $t_2$, after which the weight accelerates its motion transferring together with the belt into a uniform motion at

$$x_3 = 2x_p - x_2. \tag{5.76}$$

Further, the weight undergoes shear in point $x_4$, and so on.

In the case sequence $x_0, x_2, x_4, \ldots$ tends to some $x_\kappa$ value different from $x_p$, then the jerky motion will become uniform and the tension force of the spring will be compensated by the friction force.

The condition for the FS generation is derived from the static friction force equality to the elastic deformation of the spring:

$$F_s(t) = kx_K. \tag{5.77}$$

If to substitute $F_s(t) = F_\infty - (F_\infty - F_0)e^{-ut}$ and $x_k = x_p + \frac{\vartheta t}{2}$ in (5.77) the condition for the FS to occur [90] will be in the form

$$\vartheta_k < \frac{2(F_\infty - F_0)u}{k}. \tag{5.78}$$

If the condition (5.78), is met, the relaxation FS will take place with unstable position of the weight. In the case the velocity of the belt exceeds $\vartheta_k$, the relaxation FS become improbable in the system, although the equilibrium iremains stable.

The oscillation amplitude in the given macrosystem is found from:

$$A = \sqrt{(x_k - x_p)^2 + \left(\frac{c\vartheta}{m}\right)^2}. \tag{5.79}$$

The total period of oscillations will be

$$T = 2\frac{x_k - x_p}{\vartheta} + \pi\sqrt{\frac{m}{k}} + 2\phi, \tag{5.80}$$

where $\phi$ is determined by the relation

$$\phi\cos\sqrt{\frac{k}{m}} = \frac{x_k - x_p}{A}. \tag{5.81}$$

The stage of the relative rest of the weight on the moving belt is expressed by the formula:

$$t_1 = \frac{F_s - F(0)}{k\vartheta}, \tag{5.82}$$

where $\frac{F_s}{k} = x$—shift from the equilibrium state till shear. For simplification we have assumed nonlinearity of the static characteristic of friction

$$F_s = F(0) + q_s t_1 \tag{5.83}$$

Coefficient $q_s$ is found by substitution of the boundary conditions at $t = 0$ $F(0) = kx_0$ and at $t = t_1$ $F_s(t_1) = kx_1$. As a result,

$$q_s = \frac{F_s(t_1) - F(0)}{t_1} = k\vartheta. \tag{5.84}$$

Consequently, the friction force during $t_1$ is determined as

$$F_s = F(0) + k\vartheta t_1 \tag{5.85}$$

Duration of the stage of relative rest $t_1$ is conditioned by the friction parameters, physico-mechanical characteristics of the rubbing materials and strength of the forming bonds. The latter reduces as the sliding velocity grows. Notice that (5.83–5.85) have sense only if $t_1 \le t_0$, since at $t_1 > t_0$ the static friction force does not change and equals to $F_\infty$.

The stage of accelerated sliding of the weight (after shear) without damping account is described by the equation

$$m\ddot{x} + F(\dot{x}) + kx = 0 \tag{5.86}$$

For a general case, the friction force versus sliding is expressed by the empirical equation

$$F(\dot{x}) = (k_1 + k_2\dot{x})e^{-k_3\dot{x}} + k_4, \tag{5.87}$$

where $k_1 \ldots k_1$—constants.

Equation (5.86) is hard to solve analytically by substitution of (5.87) just as to determine coefficients $k_1 \ldots k_1$. Therefore, the duration and character of motion at this stage are described graphically using the Lienar's construction and further grapho-analytical integration of the phase trajectory [8, 12]. This procedure for calculations of FS consists of the following stages:

(1) substitution of the variables brings (5.86) to the form $\frac{dv}{dx} = \frac{-F_k(v)-x}{\vartheta}$;
(2) graph $x = -F_k(\vartheta)$ is constructed based on experimental data;
(3) the phase trajectory is constructed by Lienar's method;
(4) the diagram of the oscillation process is constructed by integrating the phase trajectory.

Above procedure has found wide application in calculations of low-frequency FS (8–10 Hz) that occur at clutch jerking. The approach proposed in [34] with account of both kinetic and static frictional characteristics has been later used by a number of researchers to calculate the FS in dry friction joints [124, 128].

An essential drawback of the equations like (5.86) is disregard of damping. The role of damping as a dissipative factor in exciting FS is very important in practice for a broad range of friction units. This approach is therefore inapplicable, e.g., in calculations of brake squeal and chatter. Afterwards, there were proposed the equations with complete consideration of the friction processes accompanied by the FS [23, 36], which, however, can be solved only by numerical methods.

Above-considered models with one degree of freedom are inapplicable for most of the real friction joints operating without lubrication because the process of frictional interaction involves at least two oscillating subsystems [129]. So, the FS are dependent on the relationship between normal oscillations inducing variations in the normal force and tangential oscillations induced in their turn by the statico-kinetic characteristics of friction [99].

Analogously, the friction units operating with a liquid LM should account for at least two degrees of freedom as they are conditioned by the rubbing elements moving along the friction path ($x$ axis) and perpendicularly to the friction surface ($y$ axis) [16, 86, 121].

The scheme in Fig. 5.18 illustrating a closed dynamic frictional system with two degrees of freedom shows that the displacement normal to the friction surface (ascending) is exercised under the action of hydrodynamic lifting force $Q$ of the lubricating layer. Notice that the phenomenon of lifting occurs in both mixed and boundary lubrication.

The generalized coordinates $x$ and $y$ are related through velocity $\dot{x}$ due to the presence of hydrodynamic effect. The feedback is exercised thanks to dependence of the friction force upon variations in the contact strain in response to surface lifting.

The linearized equations for the perturbed motion of the system are of the kind [121]:

$$\begin{cases} m\ddot{x} + c_x\dot{x} + k_x x = -k_c y \\ m\ddot{y} + c_y\dot{y} + k_y y = -k_{\text{жс}}\dot{x}, \end{cases} \tag{5.88}$$

where the members of the system of equations express the following variable components: $m\ddot{x}$ and $m\ddot{y}$—inertia forces, $c_x\dot{x}$—forces of viscous resistance, $c_y\dot{y}$—resistance forces to floating, $k_x x$ and $k_y y$—elastic forces, $k_c y$—friction forces, $k_l\dot{x}$—hydrodynamic force of lifting. Here, $c_x$ and $c_y$ are damping factors along $x$ and

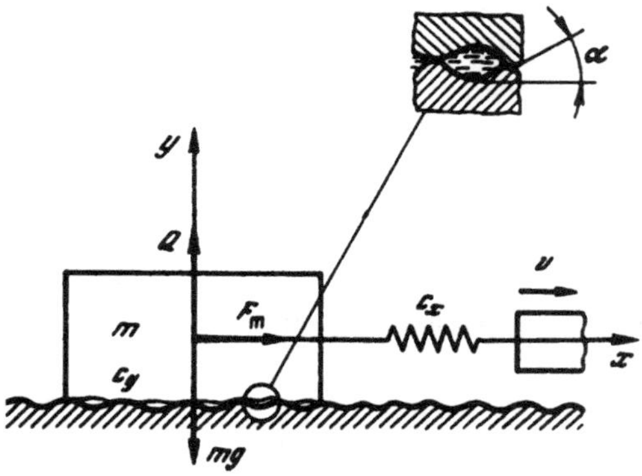


**Fig. 5.18** Frictional system with two degrees of freedom [85]

$y$ axes, $k_x$—stiffness of the unit keeping the friction element longitudinally, $k_y$—contact rigidity, $k_d = f_d k_y$—coefficient of dry friction, $k_l \approx (\mu_c A_r B^2) / n(h_0 + y_0 + B)^3$, $B$ and $n$—width of sliding guides and number of their active faces.

Stability of the system is found from the next inequality

$$k_y\left(1 - \frac{k_l f_d}{c_x}\right)\left(k_y m + c_x c_y\right) + k_x c_y^2 > 0, \tag{5.89}$$

which is true under the condition.

$$\frac{k_l f_d}{c_x} < 1. \tag{5.90}$$

Inequality (5.90) indicates that it is possible to raise stability of the system only by reducing the coefficient of dry friction and increasing damping in sliding direction.

Experimental research of the normal load and friction force spectra at boundary lubrication and dry friction has shown that fluctuations of the normal load and friction force either coincide or are close in phase, while their spectra are in fact similar [117].

Analogous results were obtained in [86] where the authors have shown that fluctuations in sliding direction and normal one are of equal frequency and amplitude dependence on velocity. This is supported by interrelation of the oscillating subsystems.

With increasing sliding velocity the relaxation oscillations (under the lowest velocities) transfer into the harmonic ones (at low velocities), and gradually cease, making the motion stable (at medium and high velocities). Similarly, with increasing stiffness of the system the amplitude lowers and oscillation frequency elevates. Increment in mass of the movable parts leads, vice versa, to impairment of stability in the system. The area of a stabile motion expands not continuously in this case but in leaps. The harmonic self-oscillations are dependent more on viscosity of the LM, e.g., increase in viscosity narrows the area of unstable motion. It should be noted that in the region of semisolid lubrication the FS are excited not only on the descending sections of the kinetic characteristic but on the ascending ones as well. Therefore, its is insufficient to estimate the tribological conditions for the SF based on the kinetic characteristic only [86].

To describe FS in the real friction joints one should often employ more complex dynamic models that take into account deformation of machine parts, presence of clearances in kinematic pairs and other factors [1]. The number of generalized coordinates defining the position of all material points of these models, i.e., the number of degrees of freedom may surpass the number of the mobility degrees. The periodic noninterruptible self-oscillations in frictional systems with $n$ degrees of freedom are analyzed in works [130, 131]. The authors of [132] have estimated stability of a bimass model with a few degrees of freedom by using a partial harmonic linearization and have proved its applicability for the real tribosystems. A design method for close to harmonic FS generated in the systems with $n$ degrees of freedom is described in [133]. Works [124, 128] have studied the mathematical models of the relaxation FS for the systems with one and two degrees of freedom.

A semi-analytical method of a harmonic balance and numeric methods are proposed in [134, 135] to study FS by a model with two degrees of freedom as applied to car transmissions. These investigations are intended to estimate the effect of variable in time factors upon stability of the system.

The modern level of the theory of FS does not afford to come to some accurate analytic solution of above-mentioned phenomena for vital practical applications. The experimental research in this respect on the real friction joints is rather expensive. The corresponding computer simulation programs have been elaborated lately performing the computation experiments on a model of the system under study. Stochastic processes make the base of computer simulation and the results are obtained in the form of statistical conclusions.

Differential equations proposed by different authors for computer simulation of FS are presented in a simplified form in Table 5.3. Equations 1 and 2 ignore damping in the system and are giving only qualitative results, being therefore inappropriate for engineering design. Equation 2 makes account for damping but neglects the effect of the static and kinetic characteristics of friction. Equations 4–6 are more substantiated, although friction-induced nonlinearity makes them unsolvable in a general case, so to solve them one should use linearization and experimental coefficients. A system of Equation 7 is more complete in terms of the processes exciting self-oscillations but is rather intricate for solution. Besides, it does not meet the conditions of self-excitement, contains the units of a similar dimensionality but different in physical sense, making it impossible to study by the methods of similarity and dimensionality.

**Table 5.3** Differential equations of FS

| Equation | Authors |
|---|---|
| 1. $m\ddot{x} + kx - F(\dot{x}) = 0$ | Den Hartog [25], Kaidanovskii and Khaikin [26, 71], Strelkov [27], Bowden and Leben [39] |
| 2. $m\ddot{x} + c\dot{x} + kx - F(\dot{x}) = 0$ | Block [44] |
| 3. $m\ddot{x} + kx - F(\dot{x}) = 0$ | Kosterin and Kragelskii [34] |
| 4. $m\ddot{x} + c\dot{x} + kx - F(\dot{x}, \ddot{x}) = 0$ | Eliyasberg [36] |
| 5. $m\ddot{x} + c\dot{x} + kx - F(t, \dot{x}) = 0$ | Deryagin et al. [22] |
| 6. $\ddot{x} + 2c\dot{x} + x - F_f(t, \dot{x}) = F_p \sin(\Omega t)$ | Duan and Singh [135] |
| 7. $\begin{cases} m\ddot{x} + c_x\dot{x} + k_x x - F(y) = 0 \\ m\ddot{y} + c_y\dot{x} + k_y x - Q(x) = 0 \end{cases}$ | Kudinov and Lisitsyn [86, 121] |

In connection with inadequacy of the mathematical models to the real processes in the friction zone as well as their intricacy for solving differential equations of motion, there arises q necessity in experimental investigations in the real or close to real environment (field tests) to obtain the empiric conditions of stability and parameters of frictional self-oscillations.

## References


1. I.V. Kragelskii, V. Gitis, *Friction-Induced Self-oscillations* (Nauka, Moscow, 1987), p. 184
2. D.V. Veinberg, G.S. Pisarenko, *Mechanical Vibrations and Their Role in Engineering* (Nauka, Moscow, 1965), p. 276
3. A.A. Kharkevich, *Self-oscillations* (Gostehizdat, Moscow, 1953), p. 170
4. Yu.I. Iorish, *Vibrometry. Measurement of Vibration and Impact. General Theory, Methods and Instrumentation* (Mashinostroenie, Moscow, 1963), p. 772
5. A. Nashir, D. Jones, D. Henderson, *Damping of Oscillations* (Mir, Moscow, 1988), p. 448
6. A.V. Chichinadze, *Design and Research of External Friction at Braking* (Nauka, Moscow, 1967), p. 231
7. F.R. Gekker, *Dynamics of Machines with Unlubricated Operation of Friction Joints* (Machinostroenie, Moscow, 1983), p. 168
8. I.V. Kragelskii, *Friction and Wear* (Mashinostroenie, Mosvow, 1962), p. 384
9. P. Wickramarachi, R. Singh, Analysis of friction-induced vibration leading to eek noise in a dry friction clutch, in *Proceedings of International Congress and Exposition on Noise Control Engineering*, Dearborn, MI, USA. August 19–21, 2002
10. S.S. Kokonin, E.I. Kramarenko, A.M. Matveenko, *Fundamentals of Design of Aviation Wheels and Brake Systems* (MAI, Moscow, 2007), p. 264
11. V.M. Shamberov, The effect of non-Coulomb dry friction on stability of automatic systems, in *Proceedings of Academy of Science*, Moscow, 2005, vol. 401(2), pp. 193–195
12. V.V. Kazakevich, G.M. Ostrovskii, To the question on oblique control in considering Coulomb friction in a sensitive element. Autom. Telemech. **17**(5), 119–214 (1956)
13. N.L. Golego, A.Ya. Alyabiev, V.V. Shevelya, *Fretting-Corosion of Metals* (Tekhnika, Kiev, 1974), p. 270
14. T. Jibiki, M. Shima, H. Akita, M. Tamura, A basic study of friction noise caused by fretting. Wear **251**, 1492–1503 (2001)
15. B.P. Gritsenko, Deformation and failure of the materials modified by plasma beams during friction. Dr. Sci. Thesis, Tomsk, 2007, p. 297
16. V.A. Kudinov, *Dynamics of Machine-Tools* (Mashinostroenie, Moscow, 1967), p. 360
17. V.A. Nalbandyan, Investigation of positioning accuracy of NC machine tools. Ind. Armenia **10**, 24–26 (1980)
18. M.I. Koval, G.A. Igonin, Comparative analysis of error fractions of NC machine tool working. Mach.-Tools Tool. **9**, 8–11 (1979)
19. V.L. Veina (ed.), *Nonlinear Problems in Dynamics and Strength of Machines* (LGU Publ., 1983), p. 336
20. Yu.A. Kashin, Some peculiarities of the high-strength wire drawing, in *Proceedings VNII Inform. Ferrous Metallurgy*, ser. 8, Hardware Production, Issue 3, Moscow, 1963, p. 6
21. I.I. Bakhman, *Vibration Mechanics* (Fizmat, Moscow, 1994), p. 363
22. B.V. Deryagin, V.E. Push, D.M. Tolstoy, The theory of sliding of solids with periodic stops (Friction-induced self-oscillations of the 1st kind). J. Tech. Phys. **26**(6), 1329–1342 (1956)
23. S. Thomas, Vibrations damped by solid friction. Philos. Mag. **9**, 329 (1930)
24. J.H. Wells, *Kinetic boundary friction*. Engineer (Gr. Brit.), vol. 147, p. 454 (1929)

25. J.P. Den Hartog, Forced vibrations with combined Coulomb and viscous friction. Trans. ASME APM **53**(9), 107–115 (1931)
26. N.L. Kaidanovskii, S.E. Khaikin, Mechanical relaxation self-scillations. J. Tech. Phys. **3**(1), 91–107 (1933)
27. S.P. Strelkov, The theory of pendulum self-oscillations. J. Tech. Phys. **3**(4), 563–572 (1933)
28. V.M. Yakovlev, On self-oscillations of a weight on a moving transportation belt. Mech. Solids **2**, 6–9 (1966)
29. B.E. Klamecki, A catastrophe theory description of stick-slip motion in sliding. Wear **101**(4), 235 (1985)
30. N.F. Bessarab, Friction-induced self-oscillations. J. Tech. Phys. **26**(1), 102–108 (1956)
31. R. Schnurmann, E. Warlow-Davies, The electrostatic component of the force of sliding friction. Proc. Phys. Soc. **54**, pt. 1(301), 14 (1942)
32. K. Shtepanek, in *Stability of motion in slide guides*, ed. by I. Tlustoy. Investigation of Metal-Cutting Machines (Mashgiz, Moscow, 1962), pp. 35–65
33. S.J. Dokos, Sliding friction under extreme pressures. J. Appl. Mech. **13**(2m), 148 (1946)
34. Yu.I. Kosterin, I.V. Kragelskii, Relaxation oscillations in elastic friction systems, in *Friction and Wear in Machines*, Moscow, vol. 11, pp. 119–143 (1958)
35. V.E. Push, *Small Displacements in Machine Tools* (Mashgiz, Moscow, 1961), p. 124
36. M.E. Eliyasberg, Design of feeding mechanisms in metal-cutting machines for smoothness and sensitivity of motion. Mach. Tools Tool. **11**, 1–7 (1951)
37. M.E. Eliyasberg, On breaking vibrations in sliding. Mach. Tools Tool. **12**, 6–9 (1951)
38. P.L. Papenhuysen, Wrijvingaproeven in verband met het Shippen van autovanden, *Ingenieur*, N. 53, S. 75 (1938)
39. F. Bowden, L. Leben, The nature of sliding and the analysis of friction. Proc. Royal Soc. Lond. Series A, Math. Phys. Sci. **169**(938), 371–391 (1939)
40. J.R. Jones, A study of stick-slip under press fit conditions. Lubric. Eng. **17**(4), 290 (1967)
41. Yu.M. Mamontova, Some investigation results of stick-slip in sliding by the asymptotic method. *The theory of Mechanisms and Machines*, issue 3 (Kharkov, 1967), pp. 87–95
42. J. Brecht, K. Schiffner, Influence of friction law on brake creep-groan. *SAE Paper* 2001-01-3138, 2001
43. S. Bahadur, The development of transfer layers and their role in polymer tribology. Wear **245**, 92–99 (2000)
44. H. Block, Fundamental mechanical aspects of boundary lubrication. *SAE Paper* 400129, 1940
45. S. Thomas, Vibrations damped by solid friction. Philos. Mag. **9**, 329 (1930)
46. N.F. Kunin, G.D. Lomakin, Soundless dry external friction of metals at low velocities. J. Tech. Phys. **24**(8), 1361–1366 (1954)
47. N.K. Myshkin, M.I. Petrokovets, A.V. Kovalev, Tribology of polymers: adhesion, friction, wear and frictional transfer (Review). J. Frict. Wear **27**(4), 429–443 (2008)
48. J.V. Strett (J. Reyleigh), *The Theory of Sound*, vol. 1 (Gostehizdat, Moscow, 1955), p. 530
49. G.A. Gorokhovskii, The method and investigation results of wear of metal-polymer friction pair elements at dry and boundary friction. Frict. Lubr. Wear Mach. Parts **2**, 43–55 (1961)
50. I.G. Nosovskii, *Influence of Gaseous Media on Wear of Metals* (Tekhnika, Kiev, 1968), p. 181
51. J.T. Burwell, E. Rabinowicz, Nature of coefficient of friction. J. Appl. Phys. **24**(2), 136 (1953)
52. D.K. Minov, The theory of cohesive force realization process at electric traction and methods of intensifying its application. *Problems in Raising Efficiency of Transportation*, issue 1 (USSR AS Publ., Moscow, 1953), pp. 7–129
53. E.E. Novikov et al., On the friction coefficient value under low sliding velocities. *The Theory and Design of Mining Machines*, 1982, pp. 39–51
54. Yu.I. Kosterin, *Mechanical Vibration at Dry Friction* (USSR AS Publ., Moscow, 1960), p. 76
55. S.S. Kedrov, *Vibrations of metal-cutting machines* (Mashinostroenie, Moscow, 1978), p. 199

56. V.D. Biderman, *The theory of mechanical vibration* (Vysshaya Shkola, Moscow, 1980), p. 404
57. G.G. Dorfman, Research and design of a heavy machine-tool part motion along slideways using simulation. Ph.D. Thesis Summary, ENIMS, 1979, p. 16
58. K. Shooter, R.H. Thomas, Friction properties of plastics. Research **2**, 533–539 (1952)
59. W.C. Milz, L.E. Sargent, Frictional characteristics of plastics. Lubr. Eng. **11**, 313–317 (1955)
60. T. Fort, Adsorption and boundary friction of polymer surfaces. J. Phys. Chem. **66**, 1136–1143 (1962)
61. N.S. White, Small oil-free bearings. J. Res. Nat. Bur Stand. **57**, 185–189 (1956)
62. D.G. Flom, N.T. Porile, Effect of temperature and high-speed sliding on the friction of Teflon sliding on Teflon. Nature **175**, 682–685 (1955)
63. D.G. Flom, N.T. Porile, Friction of Teflon sliding on Teflon. J. Appl. Phys. **26**, 1080–1092 (1955)
64. G.M. Bartenev, V.V. Lavrentiev, *Friction and Wear of Polymers* (Khimiya, Leningrad, 1972), p. 240
65. A. Schallamach, The load dependence of rubber friction. Proc. Phys. Soc. **65B**, 658–661 (1952)
66. L. Gümbel, Reibug und Schmierung im Maschinenbau. B.: Fichtenau. 321 S, 1925
67. L. Leloup, Etude d'un régime de lubrification: le frottement onctuent des paliers lisses. Rev. univ. mines. **90**(10), 373 (1947)
68. G. Tränkner, Reibundsmessungen an kleinen Langern im Gebiet der Grenzreibundsforschung auf dem Gebiete. Ingenierwesens **14**(1), 26 (1943)
69. T.M. Birchall, A.G. Moore, Friction and lubrication of machine tool slideways. Machinery **93**(2395), 29 (1958)
70. M.E. Salama, Effect of macroroughness on performance of parallel thrust bearings. J. Appl. Mech. **163**(2), 49 (1950)
71. N.L. Kaidanovskii, The origin of mechanical self-vibrations generated at dry friction. J. Tech. Phys. **19**(9), 985–996 (1949)
72. D.M. Tolstoi, P. Bio-Yao, On the friction force jump at stoppage. Proc. USSR AS **114**(6), 1231–1234 (1957)
73. V.A. Kudinov, D.M. Tolstoi, Friction and oscillation. Friction, wear and lubrication. Ref. Book in 2 vols., vol. 2, ed. by U.V. Kragelskii, V.V. Alisin (Nauka, Moscow, 1979), pp. 11–22
74. Le Suny Any, Mechanical relaxation oscillations. Proc. USSR AS, MTT, **2**, 47–50 (1973)
75. A.V. Chichinadze, A.S. Tomish, A dynamic test method for bearing materials at sign-varying friction. The methods of testing and evaluation of service characteristics of materials for journal bearings. Moscow, 1972, pp. 41–44
76. H. Schindler, Einfluss der Werkstoffpaarung auf die Gleit und Genauigkeitseigenschaften von Geradführungen. *Maschinenbautechnik*, 1969. Bd. 18, N. 7. – S. 345
77. E. Rabinowicz, The intrinsic variables affecting the stick-slip process. Proc. Phys. Soc. **71** (460), pt 4, 668 (1951)
78. F. Morgan, M. Muskat, D.W. Reed, J.B. Sampson, Friction behavior during the slip position of the stick-slip process. J. Appl. Phys. **14**(12), 689 (1943)
79. V.L.M. Veitsm, M.S. Bundur, V.E. Khitric, V.A. Shakov, The analysis of regularities in formation of dynamic characteristics of friction interacting with an elastic medium. J. Frict. Wear **6**(4), 653–660 (1985)
80. V.D. Tolstoi, New instruments for examination of friction-induced self-oscillations. Mach. Tools Tool. **1**, 22–24 (1961)
81. R. Bell, M. Buderkinm, Dynamic behavior of plain slideways. Ibid. 196. **181**(8), pt. 1, 169 (1967)
82. A.V. Chichinadze, O.S. Tenish, On the choice of materials for friction damping. Mach. Sci. **3**, 102–105 (1970)

83. I.T. Chernyavsky, O.V. Temish, *To the question on PC-aided modeling of external (dry) friction*, ed. by N.G. Brusevich. Automation of Research in Mechanical Engineering and Instrument-Making (Novosti Press Publishing House, Moscow, 1971), pp. 177–183
84. S.A. Brokly, P.L. Ko, Friction-induced quasi-harmonic oscillations. Probl. Frict. Lubr. **92**(4), 15–21 (1970)
85. B.M. Belgaumkar, The influence of Coulomb, viscous and acceleration-dependent terms of kinetic friction on the critical velocity of stick-slip motion. Ibid **12**(2), 107 (1981)
86. V.A. Kudinov, N.M. Lisitsyn, Key factors affecting smoothness of the motion of the machine platens and supports during sheared sliding. Mach. Tools Tool. **2**, 1–5 (1962)
87. J.S. Rankin, The elastic Range of friction. Phil. Magaz. **2**(10), 806–816 (1926)
88. M. Hunter, Static and sliding friction of pivot bearing. Engineering **157**(4074–4075), 117–138 (1944)
89. I.V. Kragelskii, The effect of the static contact duration on the friction force value. J. Tech. Phys. **14**(45), 272 (1944)
90. A.Yu. Ishlinskii, I.V. Kragelskii, On the friction-induced leaps. J. Tech. Phys. **14**(45), 276–282 (1944)
91. V.P. Shishokin, The effect of loading time on hardness of metals and alloys. Tech. Phys. **8**, 18 (1938)
92. I.V. Kragelskii, On the friction of unlubricated surfaces, in Proceedings of 1st All-Union Conference on Friction and Wear in Machines, USSR AS Publ., vol. 1, 1939
93. V.S. Shchedrov, Investigation of friction and wear processes on a sliding contact in machines. Dr. Techn. Sci. Thesis summary, IMASh, Moscow, 1953, p. 19
94. P.G. Howe, D.P. Benton, I.E. Puddington, London-Van-der-Waal's attractive forces between glass surfaces. Can. J. Chem. **33**, 1375 (1955)
95. E. Rabinowicz, The nature of static and kinetic coefficients of friction. J. Appl. Phys. **222**(2), 1373 (1951)
96. N.B. Demkin, *Contact of Rough Surfaces* (Nauka, Moscow, 1970), p. 228
97. N.F. Kunin, G.D. Lomakin, On interrelation between static and kinetic friction. J. Tech. Phys. **24**(8), 1367–1370 (1954)
98. F.P. Bowden, D. Tabor, *Friction and Lubrication of Solid Bodies* (Mashinostroenie, Moscow, 1968), p. 543
99. I.P. Kornauli, *On the Leaps During Friction* (Mechanics of Machines, Tbilisi, 1981), pp. 86–92
100. A.V. Chichinadze, O.S. Temish, Friction damper design. News. Mech. Eng. **1**, 12–14 (1971)
101. A.P. Amosov, Relaxation oscillations in external friction. Proc. USSR AS **212**(3), 569–572 (1973)
102. A.P. Amosov, On the conditions generating external friction oscillations. Mach. Sci. **5**, 82–89 (1975)
103. V.P. Sergienko, S.V. Bukharov, A.V. Kupreev, Tribological processes on contact surfaces in oil-cooled friction pairs. Proc. NAS of Belarus. **51**(4), 86–89 (2007)
104. V.I. Kolesnikov, V.P. Sergienko, S.N. Bukharov et al., Investigation of noise and vibration in tribopairs of railway rolling stick by acoustic interferometry and laser vibrometry. Bull. Rostov State Univ.Commun. Lines **35**(3), 5–9 (2009)
105. V.I. Kolesnikov, V.P. Sergienko, V.V. Zhuk, V.A. Savonchik, S.N. Bukharov, Friction joints: investigation of tribological phenomena in nonstationary processes and some optimizing solutions, in *Proceedings 7-th International Symposium on Friction Products and Materials*, September 9–11, 2008, Yaroslavl, pp. 25–33
106. A.I. Sviridenok, S.A. Chizhik, M.I. Petrokovets, *Mechanics of the Discrete Friction Contact* (Science and Technique, Minsk, 1990), p. 272
107. M. Nishiwaki et al., A study on friction materials for brake squeal reduction by nanotechnology. *SAE paper* 2008-01-2581, 2008
108. Ostermeyer G.P., Müller M. New developments of friction models in brake systems // *SAE Paper* 2005-01-3942, 2005

109. G.P. Ostermeyer, M. Müller, H. Abendroth, B. Wernitz, Surface topography and wear dynamics of brake pads. *SAE Paper* 2006-01-3202, 2006
110. M. Müller, G.P. Ostermeyer, A cellular automaton model to describe the three-dimensional friction and wear mechanism of brake systems. Wear **263**(7), 1175–1188 (2007)
111. G.P. Ostermeyer, On tangential friction induced vibrations in brake systems. *SAE Paper* 2008-01-2850, 2008
112. S.N. Bukharov, Reduction of vibroacoustic activity of metal-polymer tribojoints in nonstationary friction processes. Summary of Ph.D. Thesis, 05.02.04, MPRI NASB, Gomel, 2010, p. 24
113. I.I. Blekhman, *Synchronization of Dynamic Systems* (Nauka, Moscow, 1971)
114. I.I. Blekhman, *Synchronization in Nature and Engineering* (Nauka, Moscow, 1981), p. 122
115. A.S. Pikovsky, B.G. Rozenblum, Yu. Kyrts, *Synchronization. A fundamental nonlinear phenomenon* (Tekhnosfera, Moscow, 2003), p. 496
116. V.P. Sergienko, S.N. Bukharov, Noise and vibration in frictional joints of machines. Tribologia **217**(1), 129–137 (2008)
117. A. Soom, C. Kim, Roughness-induced dynamic loading at dry and boundary-lubricated sliding contacts. J. Lubr. Tech. **105**(4), 75 (1983)
118. D.M. Tolstoy, Self-excited vibrations of a slider dependent upon its rigidity and their effect on friction. Proc. of USSR AS **153**(4), 820–823 (1957)
119. R. Courtel, Sur l'observation des certains dommages périodiques causes aux surfaces par le frottement et leur interpretation. C.r. Acad. Sci. **253**, 1758 (1961)
120. D.M. Tolstoy, R.L. Kaplan, To the problem on the role of normal displacements at external friction, ed. by I.V. Kragelskii. New in the theory of friction (Nauka, Moscow, 1966), pp. 42–59
121. V.A. Kudinov, Vibration in machine-tools. Vibration in Engineering. Ref. Book in 6 vol., vol. 3, ed. by F.M. Dimentberg, K.S. Kolesnikov (Mashinostroenie, Moscow, 1980), pp. 118–130
122. B.V. Budanov, V.A. Kudinov, D.M. Tolstoy, Interrelation of friction and vibrtation. J. Frict. Wear **1**(1), 15–21 (1980)
123. A.I. Sviridenok, N.K. Myshkin, T.F. Kalmykova, O.V. Kholodilov, *Acoustic Electrical Methods in Triboengineering* (Allerton Press Inc., New York, 1988)
124. V.F. Petrov, On the mechanical self-vibrations at dry friction of the systems with one degree of freedom. Bull. of MGU, Ser. 1, Math. Mech. **2**, 86–92 (1967)
125. J.D. Ferry, *Viscoelastic Properties of Polymers* (Wiley, New York, 1980), p. 662
126. Yu.M. Pleskachevsky, E.I. Starovoitov, A.V. Yarovaya, *Dynamics of Metal-Polymer Systems* (Belarus Science Publ., Minsk, 2004), p. 386
127. V.P. Sergienko, S.N. Bukharov, Vibroacoustic activity of tribopairs depending on dynamic characteristics of their materials. Mech. Mach. Mech. Mater. **9**(4), 27–33 (2009)
128. S.V. Baev, To the question on relaxation oscillations in the systems with dry friction, in *Proceedings of Dnepropetrovsk Institute of Railroad Transport, "The theory of oscillations and dynamics of bridges"*, issue 89, Kiev, 1969, pp. 33–40
129. J.A.C. Martins, J.T. Oden, F.M.F. Simoes, Recent advances in engineering science: a study of static and kinetic friction. Int. J. Eng. Sci. **28**(1), 29–92 (1990)
130. L.P. Pavkenko, V.B. Golubev, To the design of friction-induced self-oscillations in transmissions of motor vehicles. Theory Mech. Mach. **31**, 57–68 (1981)
131. N.D. Salnikova, To the question of friction-induced self-oscillations in the systems with a finite number of degrees of freedom. Izv. VUZov, Mashinostroenie **6**, 54–59 (1968)
132. S.V. Belokobylsky, R.F. Nagaev, *The Method of Partial Harmonic Linearization in the Problems of Friction-Induced Self-Oscillations in Mechanical Systems with a Few Degrees of Freedom*, vol. 5 (Mashinostroenie, Moscowm, 1985), pp. 27–31
133. V.O. Kononenko, Self-vibrations in mechanical systems induced by friction. Abstr. Dr.Sci. Thesis, Inst. Constr. Mechanics of Ukrainian AS, Kiev, 1953, p. 16

134. C. Duan, R. Singh, Influence of harmonically-varying normal load on steady state behavior of a 2DOF torsional system with dry friction. J. Sound Vib. **294**, 503–528 (2006)
135. C. Duan, R. Singh, Forced vibrations of a torsional oscillator with Coulomb friction under a periodically varying normal load. J. Sound Vib. **325**, 499–506 (2009)
136. S.A. Brokly, N. Davis, Time dependence of static friction. Probl. Frict. Lubr. **1**, 57–67 (1968)

# Chapter 6
# Noise and Vibration in Nonstationary Friction Processes

The present chapter deals with nonstationary friction processes typical for automotive brakes and transmissions. A review is presented concerning the theoretical and experimental research of noise and vibration in brakes and transmissions of mobile vehicles along with the frequency and phenomenological classifications. The mechanisms of noise and vibration generation in the friction systems are described. The analytic, numeric and experiment-calculated investigation methods are considered in view of obtaining adequate design models. The analysis is given of the advanced experimental methods and the results are presented in forecasting vibroacoustic activity of tribopairs based on design methods. The authors also touch upon the basic approaches to abating noise and vibration in brakes.

## 6.1 The Main Types of Nonstationary Friction Joints

Operation of the rubbing joints at nonstationary friction is characterized by significant variations in friction conditions on the contact surfaces. These conditions include the velocity, loading, and temperature, as well as physico-mechanical, friction and wear properties of the rubbing bodies [1]. The friction is considered to be nonstationary if at least one of above-named parameters influencing conditions on the friction contact changes in time.

The examples of most spread nonstationary friction joints are the brake systems and friction clutches. Operation of the brake units and transmissions is interrelated with the friction forces generated by the rubbing bodies. The brake systems are intended to suppress the kinetic energy of the rotating or reciprocating masses. The relative sliding velocity may be in this case reduced to zero (pull up) or till some desired value (slow-down). A friction clutch is used as a rule to speed-up a motionless or a moving at some speed mass till a given one [2–4]. Operation of such joints is characterized by variations in all parameters of the friction process and conditions on the friction contact, i.e., the contour and actual contact areas, and the contact spot size.



V.P. Sergienko and S.N. Bukharov, *Noise and Vibration in Friction Systems*,
Springer Series in Materials Science 212, DOI 10.1007/978-3-319-11334-0_6

The joints of the quasi-stationary friction include the chain and the belt transmissions, rolling and sliding bearings at rubbing over a fresh track [5, 6], friction variators, gearing and so on [7, 8]. After a prolonged operation under invariable load and velocity parameters a quasi-stationary state may appear in the joints in which the constant volume temperature conditions are interfered by the considerable temperature deviations on some contact sites [9–11].

Despite above-named common features of nonstationary processes, the conditions on the friction contact, as well as the occurring phenomena may differ significantly even in the one-type friction joints depending on the tribopair materials, design, operation regimes, presence of the lubricants and their properties. Hereinbelow, we present a review of the operation peculiarities of the brakes and friction clutches adopted today in engineering spheres.

### 6.1.1 Brake Mechanisms

The arrangement of the brake system is in principle realized in the next embodiments: the belt, shoe, disc, and track brakes (Fig. 6.1) [1].

*The belt brakes* (Fig. 6.1a) are commonly applied in the tractor and agricultural machinery, technological equipment (drilling rig winches) and other devices. The frictional materials used in the belt brakes are typically rubber or resin-based polymer composites combined with a binder [1, 12].

Different-purpose belt brakes are intended for the 0.3–1.0 MPa contact pressures, 1–20 m/s initial sliding velocities, and the mean surface temperature varying between 800–900 °C. The belt brakes may operate with/without a lubricant. The operation regime is, as a rule, rotary/short-term. Their peculiarity is in a non-uniform loading of the shoes that depends upon the brake design and the transmitted torque value. This non-uniformity of the shoe loading reduces during operation after the run-in period.

*The shoe (block) brakes* (Fig. 6.1b, c) are intensively used in practically all types of vehicles, as well as technological equipment and handling machinery [13–15]. Thanks to a broader field of application, diverse environments and service conditions, shoe brakes employ a wider range of frictional materials as compared to the belt types. The contact pressure range in these brake arrangements is within 0.3–1.5 MPa, velocities till 50 m/s, and temperatures 100–1,000 °C [16]. The shoe brakes may be arranged as a drum-type that contacts either the internal or external generatrix of the drum with the coefficient of mutual overlap $k_{mo}$ in the range $0.2 < k_{mo} < 1$; and a drum chamber type ($k_{mo} \approx 1$). The former are commonly two-shoe designs with a mechanical, hydraulic, pneumatic or electromagnetic lever drive [2, 13]. In addition, these brakes are characterized by a non-uniformity of load distribution between the shoes or over a single shoe caused by the friction torque, as well as rigidity of the very shoe because of its large size (60–90°) on the arc and the large lever drive units [17, 18].

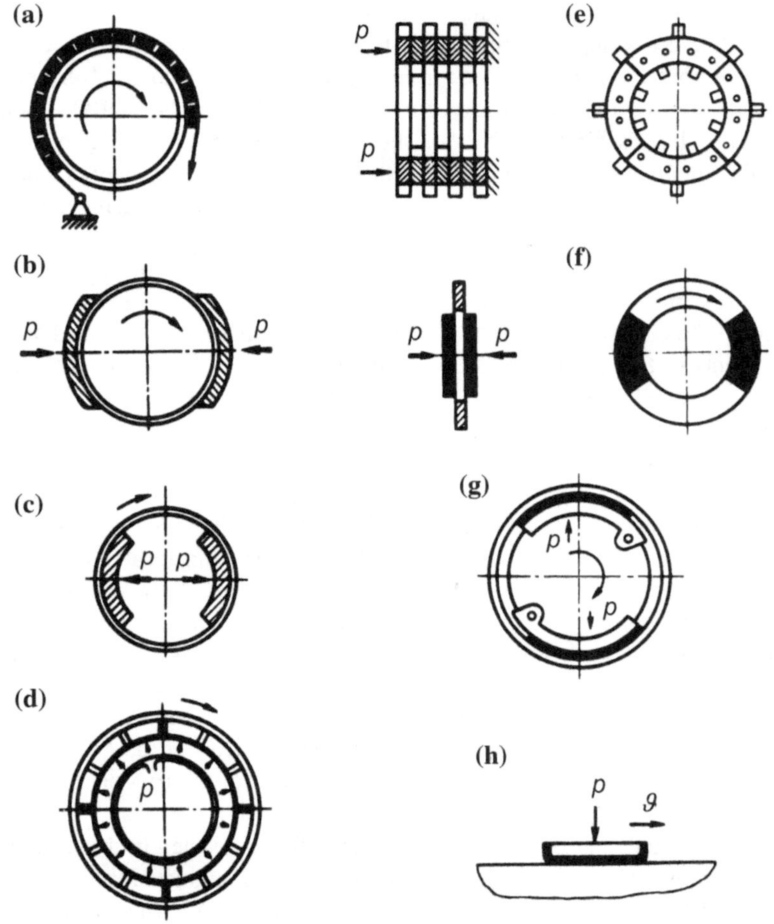


**Fig. 6.1** The types of brake designs: **a** belt brake; **b** drum-type shoe brake with external shoe mounting; **c** drum-type shoe brake with internal shoe mounting; **d** chamber shoe brake; **e** multidisc brake ($0.5 \le k_{mo} \le 1$); **f** disc-block brake ($k_{mo} < 0.5$); **g** centrifugal block brake; **h** track brake

*The chamber or air-actuated brakes* (Fig. 6.1d) show certain advantages over the two-shoe ones thanks to a more even load distribution across the nominal contact area. This reduces wear inhomgeneity on the shoe and the tribopair as a whole, and temperature fluctuations on the drum [13, 19]. Above-named brakes are fast-responding and convenient in handling. However, their design and maintenance complexity, and the necessity in a compressed air source are reducing their sales appeal.

*The disc and multidisc brakes*. The main advantage of these brakes in contrast to the shoe ones is compactness, and rather high friction torques despite their small size, manufacturing and adjustment simplicity of the friction elements, technological effectiveness and loading uniformity of the conjugated units.

Above features extend the range of fictional materials used for the disc brakes. This also concerns application of alike tribopairs for the lubricated friction purposes. The contact pressure ranges from 1.0 to 3.0 MPa, velocity 1–50 m/s, and temperature on the friction surface of the loaded brakes reaches 1,100–1,300 °C.

The disc brakes similarly to the shoe ones can be fit with a mechanical, hydraulic, pneumatic or electromagnetic drive [1, 19, 20]. They may operate with/without a lubricant in various gaseous media, in a single-time or the repetitive short-time modes.

The tendency to adopt more powerful friction systems in the modern mobile vehicles has brought to creation of the novel friction joint designs, in which the dissipative processes occur in the liquid media—oil on the main. The major requirements to friction materials operating in the oil medium include the reduced heat loads on the friction pairs and, consequently, a lowered wear rate, buckling of the rubbing components, prevention of the impact loads on the transmission members and brakes [21, 22]. The dry friction mode is substituted by the fluid or boundary processes. Besides, the forced heat removal from the friction zone stimulates improvement in the capacity, life and durability of the friction joints. This inevitably reduces the friction force, which can be compensated by adding the number of friction pairs [23]. A multidisc brake consists as a rule of $n = 2$–$6$ movable discs (rotor) and $(n + 1)$ immovable discs (stator) making together $2n$ friction planes [24]. A smoother application of the brake is attained by reducing the difference between the static and kinetic friction coefficients of the materials. It is peculiar for the frictional interactions in oil to weaken both the hydraulic and rheodynamic properties of the lubricating materials typical for their volume phase, and to strengthen the effect on tribological parameters of the friction elements and counterbodies [25, 26].

The discs of the loaded multidisc brakes are often subjected to the temperature-induced warping. This reduces the contour and nominal contact areas, increases the local temperature loads and local wear of the discs [1, 27, 28]. These brakes display a reduced brake torque due to the compressing axial loads on the discs arising from the friction-induced losses in the splines [29, 30]. To make heating of the discs more even, their rigidity should be lowered. With this aim, the friction elements are made multicomponent consisting of separate sectors able to self-center and ensure a uniform load distribution over the nominal contact. The perfected reliability and service life of the oil-cooled friction joints may be reached through solution of such tasks as stability of the oil film on the friction contact. In this connection, it is important to understand the mechanisms of tribological phenomena observed on the friction surfaces of the oil-cooled tribojoints [8, 31].

*The centrifugal brakes* (Fig. 6.1g) are mainly used in speed governors and other control mechanisms. The centrifugal forces operating directly on the friction elements or via a system of transfer mechanisms change the load on the friction contact. These brakes are difficult to control, therefore the fiction elements are wearing hard and display operation instability.

*The track brakes* (Fig. 6.1h) have found wide application in the railroad transport as a skate and magnetic track brakes. Their peculiarity is to rub constantly against a fresh track, i.e., the sliding support (brake trig) is constantly contacting the counterbody (rail), which means that new areas are brought into contact all the time.

This type of brakes has acquired a distinctive mechanism making it independent of the rail-wheel cohesion. These brakes are extensively used in the express trains and switchers [9, 32]. Being used together with the shoe wheel brakes, they nay increase the braking efficiency by 30–40 %.

The electromagnetic track brakes turn to be most usable [5, 32] thanks to their successful operation at high speeds, specific friction power and frequent emergency braking. Their friction linings can be made of cermet, steel st.2, st.3, graphitized cast iron or the like materials [16, 33] that are more resistant to wear than the polymeric ones, and undergo less wear than the contacting rails. The surface temperature of the electromagnetic track brakes of the rolling-stock may reach 1,000 °C and more when the initial speed of braking reaches 150–200 km/h.

It is to be noted that pantographs of the electric trains follow the geometry of a sliding support. They operate in a quasi-stationary mode at a constant speed, and transfer to a nonstationary regime at acceleration or braking of the electric train [9, 11, 34].

### 6.1.2 Friction Clutch

The disc and air-actuated clutches are extensively used similarly to centrifugal and drum types in which the pressing force of the shoe-drum engagement is created by the centrifugal forces. The pressure, temperature and speed ranges are roughly the same as in other similar brake designs. The performance of a clutch is purpose-oriented, i.e., fast or slow acceleration, change of speed, engagement–disengagement of the drive, etc. Since the clutch is a connecting link between the driving and driven parts (Fig. 6.2), its behavior at skidding depends not only on the friction properties of the tribopair, contact pressure, speed and design, but also on the characteristics of the drive, and inertia moments of the driving and driven elements.

It is characteristic for the clutches to operate in a recurrent short-term mode. For instance, the automotive clutches of different classes and purpose may operate at 5–20 actuations/h, while in the forging and press equipment this figure reaches 1,000.

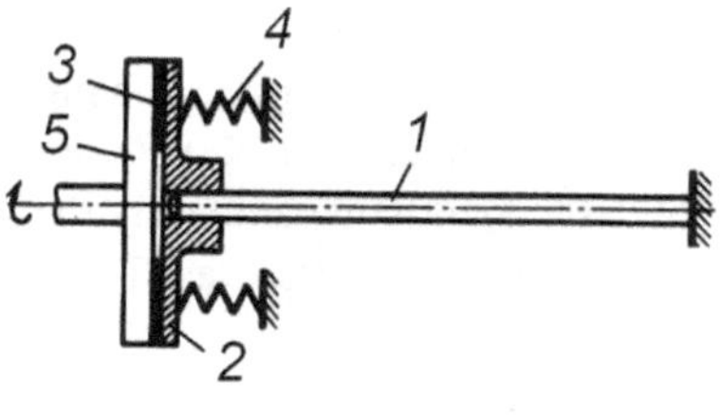


**Fig. 6.2** Scheme of a friction clutch: *1* transmission; *2* driven disc; *3* friction lining; *4* spring; *5* flywheel

Skidding of the automotive clutches may last till 3 s, for the forging equipment and machine-tools it is 0.1–0.5 s [1, 13, 14].

The air-actuated clutches are used in the drive systems of drilling rigs in which skidding lasts about 0.2–0.8 s [35].

In some types of aggregates time of skidding of the clutch lasts long, wherefore they need slow acceleration, e.g., in powerful separators of foodstuff equipment fit with a centrifugal clutch, where skidding takes 5–6 min [1].

A distinguishing feature of the majority of clutches is that their time of actuation is commensurable to skidding. It means that skidding takes place chiefly at a variable and constantly growing pressure on the friction contact. In contrast to braking, which greater or lesser share occurs under a constant pressure, at skidding it varies and affects thereby all varying parameters of the process. The changes in these parameters are dependent on the friction and wear characteristics of the tribopair.

Despite a continuous work on perfection of the composition, properties of the friction materials and designs of above-mentioned joints of nonstationary friction, the problems of raising the efficiency and reducing impact loads in transmissions and braking units still remain to be solved [36–40]. Of no less importance seems to be the factor of improving vibroacoustic parameters of friction joints. Elevated noise and vibration levels in transmissions and brakes are connected with adverse tribological phenomena that impair durability of machines [41]. This fact together with inferior subjective perception of the vehicle quality poses a task to predict and make account for the vibroacoustic characteristics of friction joints at the stage of designing [38, 42].

## 6.2 Noise and Vibration in Brake Systems

The brake systems are nonlinear mechanical devices with a probable dynamic instability as their essential feature. This means that brakes may display several probable vibration modes along with perturbation affects under preset parameters of the system [43–45]. Transfer from one stable state to another is accompanied by an abrupt change in the vibration amplitude of the system. In this connection, it is important to estimate their occurrence in a system, to isolate the states actually realized, and predict their probability in the real friction joints. It is critical since their presence in a system is often accompanied by a high-frequency acoustic radiation. For instance, *squeal* is a noise generated at a high acoustic pressure at one or several discrete frequencies within above 1,000 Hz level, i.e., with an expressed tone character [38, 46]. Squeal is mainly generated by the metallic drum disc when its HF flexural vibrations are exciting sound waves. The vibrations below 1,000 Hz may produce moan or groan generated by a combination of the brake system elements, as well as by the car body or suspension members.

### *6.2.1 The Factors Influencing Noise and Vibration in Brakes*

Dynamic instability of a brake system may be caused by a combination of factors. From the viewpoint of the methods used to eliminate this instability, they can be subdivided into two chief groups of tribological and structural factors [47, 48]. The tribological factors may include instability of the friction forces on the interface between solids due to the friction-induced relaxation self-oscillations [49, 50], or geometrical imperfections of the rubbing surfaces [51, 52], as well as the friction coefficient dependence on the contact pressure or its distribution on the friction surface [53, 54]. Besides, the negative gradient of the friction coefficient dependence of sliding velocity may also be one of the factors (the falling kinetic characteristic of friction) [47, 55, 56]. The structural factors, such as geometry, elastic and damping characteristics of the braking system, internal and external links defining the dynamic properties of the brakes present a cooperative momentum of all its elements. The dynamic instability induced by the structural factors is attributed mainly to the modal coupling [57–59]. In practice, it is justifiable to account for the dominating in a given situation mechanism.

It is accustomed to consider the NVH characteristic of the brake units (*noise*, *vibration*, and *harshness*) as a term reflecting the intensity of noise and vibration. It is dependent on a combined interaction of the car suspension and the elements of the brake system (friction pad, brake disc, support, and etc.). Figure 6.3 illustrates the main factors affecting NVH and characterizing a braking unit [60].

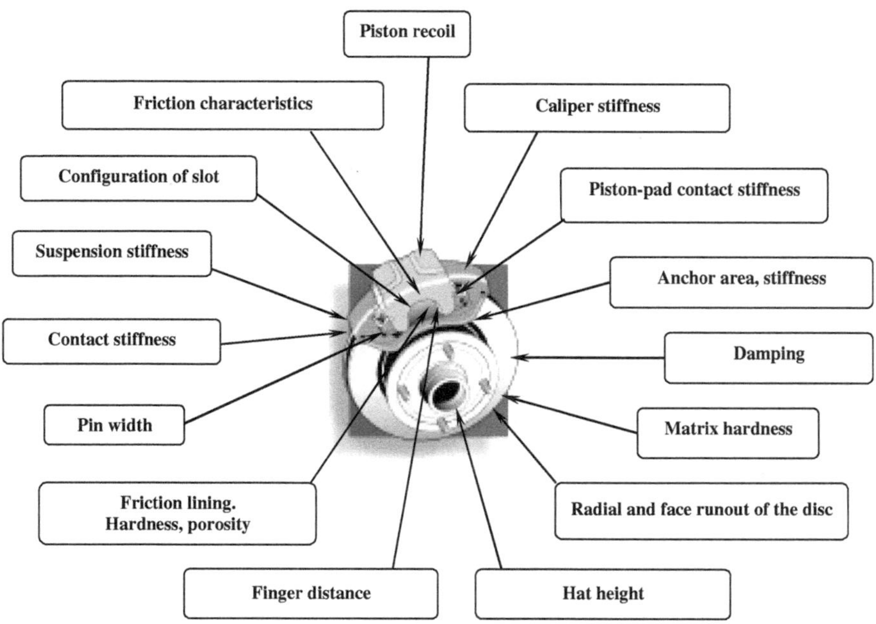


**Fig. 6.3** The factors effecting NVH characteristics of brake units

### 6.2.2 Classification and Physical Characteristics of Vibroacoustic Effects on the Friction Contact

*Frequency classification.* Noise and vibration in brakes is traditionally classified in terms of the frequencies dominating in these two phenomena. A classification generally accepted for the brake noise is illustrated in Fig. 6.4 [61]. According to this classification the oscillation frequencies found below some definite threshold (100, 500 or 1,000 Hz) belong to the LF vibration. Vibroacoustic phenomena in the frequencies above the mentioned threshold are called the HF noise, which includes squeal. We differentiate between the following types of vibration and noise generated at nonstationary friction: *squeal* [50, 59], *moan* [62, 63], *groan* [55, 64, 65], *cold judder* due to disc thickness variations and *hot judder* arising from the heat-induced stress strains [66, 67].

There are two different in principle types of vibrations in the LF range. These are forced vibrations or judder and groan, as well as interrelated with them noise called hum and moan, correspondingly. The frequency of the forced vibration is usually lower than that of the groan, which in its turn is of a lower than squeak. However, in practice, the frequency bands of judder and groan are overlapping in the range 400–500 Hz. In any case, judder is easily identified since its frequency is proportional to the car speed, whereas squeak is independent of speed.

In contrast to groan that appears in response to a certain type of staticokinetic frictional characteristics, judder is caused by changes in the friction force different from self-excited oscillations. These may be forced oscillations of the friction force because of imperfections of the friction surfaces of the tribopair (due to beating, uneven wear, transfer films or temperature instability), as well as non-uniformity of the friction properties over the friction area of the brake disc. From the other hand, squeak is brought about by dynamic instability of the brake system and is related to the resonance characteristics of the brake elements and mode coupling.

The main drawback of above frequency classification is that the phenomena of one and the same physical origin but different frequencies may be related to

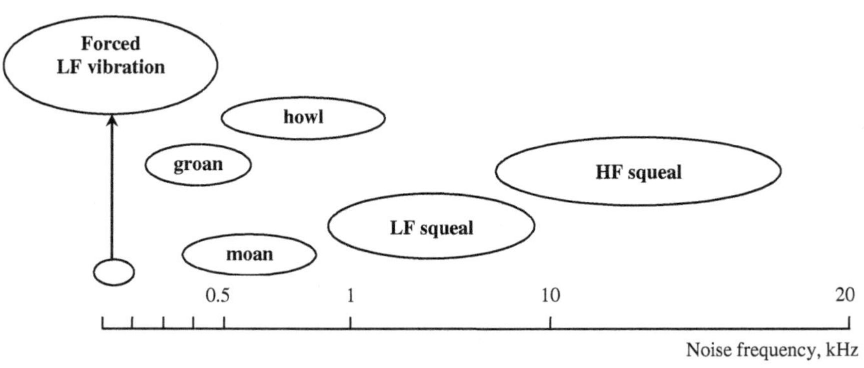


**Fig. 6.4** Frequency range of different vibroacoustic effects generated by automotive brake joints

different types. From the other hand, fundamentally different phenomena can termed identically. Nevertheless, this classification reflects subjectively the perception of these phenomena by the driver and the passengers.

*Phenomenological classification.* It is based on the physical origin of the phenomena exciting vibroacoustic activity of the brake joints was proposed by Jacobson in [68]:

1. *Forced vibration.* Forced vibrations are represented by judder and related structural noise called hum. Both cold and hot judder we relate to the LF range (5–60 Hz) according to the frequency classification. In both cases we assume that the dominating factor in generating forced oscillations of the braking torque and vibrations in the brake system is geometrical imperfection of the contact surfaces at the macroscopic level. Friction-induced heat generation in brakes and transmissions of mobile vehicles is the cause of thermoelastic deformation (warping) of contacting bodies, that affect pressure distribution on the friction contact. When the sliding velocity is rather high in conditions of uneven non-stationary heating, the situation results in thermoelastic instability or the appearance of so-called "hot spots". This leads to the LF fluctuations of the friction torque called "*hot judder*". Uneven wear of the tribopairs resulted from thermoelastic phenomena or run-out of the metal disc at actuation leads to oscillations of the braking torque perceived inside the car in the form of beating of the control elements (steering wheel, brake pedal) and chatter of the interior, which is termed as *cold judder*.
2. *Friction-excited self-oscillations.* We relate groan and connected with it noise termed moan to the group of self-oscillations. Groan results from frictional instability occurred under certain types of friction coefficient dependence on sliding velocity. It is known as a "negative damping" [55]. Groan appears in the brakes when the brake is being gradually released simultaneously with application of the torque on the wheel. The pressure on the brake block drops, so the wheel torque exceeds the braking force moment, imparting thereby discontinuity to rotation of the wheel and skidding. As a result, the wheel may stop unless an additional twist torque or further pressure drop on the lining is initiated. The repeated cycles of the stick-slip motion might lead to a strong and lasting vibration (till the car stop) not only of the braking system but other elements of suspension, body, control members and interior parts as well [69, 70]. In contrast to the cold and hot judder, the frequency of groan does not depend on rotation velocity of the wheel and is found within 30–600 Hz [61, 71]. It is characteristic for groan to display a great number of higher harmonics in its vibration spectrum. For instance, a car suspension with a McFerson's strut, is characterized by the main and usually the first harmonic corresponding to the stick-slip motion that appears in the range 20–50 Hz [70]. The higher harmonics induce the wideband noise generation in the cabin. It occupies, as a rule, higher levels but in contrast to squeal, it is hard to identify as it bears a latent character [72, 73]. Another form of the LF noise with the frequency range similar to groan is moan which frequency ranges between

100–1,000 Hz. The difference between moan and groan is that the former appears during a uniform motion with a slow down (by applying constant pressure on the brake block). In both cases the brake experiences jerking vibration accompanied by a noise. In view of its unexpected impact on the man, moan is a highly undesirable phenomenon. It should be noted that moan in contrast to groan never appears as a structural vibration [70]. Nevertheless, vibration of the brake elements together with the car body and suspension may generate moan. It may be considered because of its origin and mechanism as a low-frequency manifestation of squeal [62, 63, 73]. When instability results from the mode coupling [57–59] leading to vibrations of the flutter kind even if the friction coefficient is ideally constant, some researchers reckon it rational to analyze the dynamic characteristics of the brake system and update the design. From the other hand, structural instability may be caused by the factors connected with the frictional process dependent mainly on the tribological properties of the rubbing bodies.

3. *Resonant oscillations*. This type of oscillations appears in the form of noise (squeal) propagating rather by air than over the car structure. *Squeal* is a most frequently occurred and studied types of the brake noise. Squeal is defined as a noise with a high sound pressure level generated on a single or a few discrete frequencies in above 1,000 Hz range, i.e., it bears an expressed tonal character [38, 46, 61]. Squeal is excited by the HF free bending vibrations of the shoes or a rotating brake disc caused by frictional microoscillations. HF squeal may be induced by the resonant phenomena due to forced vibrations of the thin-walled brake elements as a result of the brake torque instability. The main source of squeal is an HF bending vibration of the metallic brake disc that generates the corresponding sound waves. To a lesser degree squeal may be a result of the brake shoe vibrations in the 4–10 kHz range [74].

It should be underlined, that subdivision of the brake noise into groan and squeal is to some extent conventional and reflects basically the methodical specifics in scientific studies of these processes.

A special classification subdividing frictional self-oscillations into the types proceeding from the kind of rupture of the bonds formed during friction on individual microcontacts is presented in works [75, 76].

1. The chaotic surface microseparations and scuffing transformed into a weak noise (friction with vibration of the 1st kind).
2. Simultaneous ordered separations of multiple microcontacts on some friction surface areas (friction with vibration of the 2nd kind).
3. Simultaneous failure of microcontacts in the moments of compete separation of the total friction surface (friction with macrovibrations of the whole body or friction with vibration of the 3rd kind).

According to above classification, groan is always related to a stick-slip sliding of the rubbing elements, i.e., vibrations of the 3rd kind. The HF noise is, as a rule, connected with the friction-induced acoustic phenomena with vibrations of the 1st and 2nd kinds. However, the experiments have proved that only in some definite cases the total area of the lining can be brought in a discontinuous sliding contact like a rigid body [72, 73, 77, 78].

## 6.3 Methods of Experimental Investigations of Noise and Vibration in Brakes

Various experimental and theoretical procedures have been elaborated to study vibration and acoustic phenomena in machine joints during operation. The experimental research involves the ride tests and the development studies. The theoretical tests include the analysis and numeric simulation of vibration and acoustic processes.

### *6.3.1 Ride Tests of Brake Systems*

To achieve the reliable and reproducible data on noise and vibration of machine units in interaction with different external factors the cars are subjected to the ride tests. The characteristics intended for further consideration of the sources, frequency, time noise and vibration and other variables are estimated in the course of the ride tests [38, 79]. Recently developed design and experimental methods make possible to differentiate between the brake noise and the external interference. It is important to find out which elements of the brake system are vibrating, which are noise emitting, and to define their spectra (frequencies and levels). It is essential to determine the effect of temperature, pressure, velocity and other factors on vibroacoustic activity of tribopairs in the brake units. Figure 6.5 presents the results of the ride tests describing all cases in which the brake noise appears as a function of the friction surface temperature.

The ride test results are more reliable but the number of controllable characteristics is restricted and insufficient for solving the optimization problems, selecting friction materials for the tribopairs or refining a tribojoint as a whole. This situation is influenced by, e.g., variations of the cohesion factor between the wheels and the road carpet in response to weather conditions, driver's style or peculiarities of riding [38, 80–82].

### *6.3.2 Development Testing*

The development or bench tests are usually based on a valid statistical description of behavior of a passenger car on the road. There are two main types of the rigs for the bench tests, namely, the inertial and drag machines [38, 83]. The kinetic energy

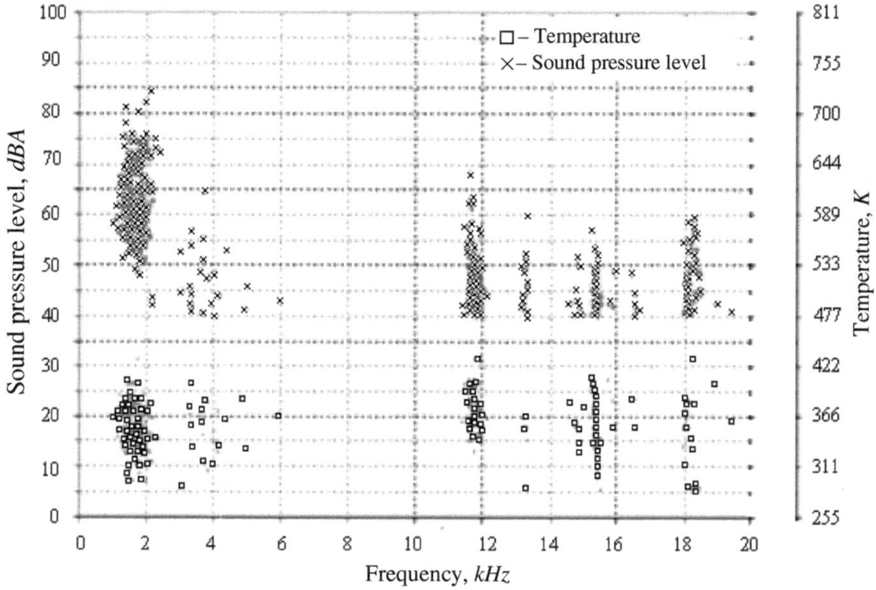


**Fig. 6.5** Ride test results for brake noise as a function of temperature [38]

of the inertial benches accumulated by the inertial mass is dissipated by one or several brake elements. The problems of the inertial benches consist in simulation of air cooling of the brake and provision of the accurate acoustic measurements. A traction bench includes a gearbox and a motor to reach a high overload capacity in terms of the driving moment. The bench maintains a preset speed and loading parameters, including simulated braking conditions. The traction bench was initially developed for vibroacoustic tests, wherefore it was commonly installed in a spacious acoustic chamber with artificial cooling. It is advantageous over the inertial bench since it enables to try the whole automotive suspension and simulate adequately cooling of the brakes [84]. A general view of the benches used for vibroacoustic testing of car brakes is presented in Figs. 6.6, 6.7 and 6.8.

The bench test programs for noise and vibration estimation in brakes are subdivided into two categories, namely: the tests following a certain program (matrix), or those simulating the real motion. The matrix programs consist of sequential cycles of brake applications, each characterized by a number of steps with a gradually rising temperature and braking pressure. The matrix usually reproduces most closely service conditions and thermal loading of the brake on the rout. The parameters like environment temperature and humidity, temperature of the disc or lining, pressure in the brake drive, rotation velocity of the wheels, acceleration rate of the car, and etc., are accurately maintained and controlled [38, 79].

Unfortunately, there is lack of unified recommendations today on the choice of the bench type and program for the vibroacoustic tests oriented on a certain brake design [36–38, 48, 79, 83, 88]. Simulation of a real motion assists in reproducing

**Fig. 6.6** A general view of a drag-type bench deigned to study the high-frequency noise (squeal) in the automotive disc brake [84]

**Fig. 6.7** A general view of a drag rig fit with suspension elements designed to study noise in the automotive drum brakes [84]

**Fig. 6.8** A general view of the inertial bench for multidisc oil-cooled (wet) brake tests simulating the real operation conditions of a mining truck brake of 75–130 tons carrying capacity [85]

road conditions on a test bench and achieving the friction, wear and thermal responses of the brake units [83]. However, the experts in the field have not come to a consensus concerning the degree of conformity between the load the car experiences during the ride tests in contrast to the bench one. When the data of the ride tests is input into the bench test matrix, one should reflect the features of the road carpet and the car motion specifics. Usually, the bench tests reproduce road conditions with insufficient accuracy, so the task their accurate simulation requires urgent solution [36–38]. Some of brake designers recommend using a SAE 2521 (USA) standard for the bench tests, although it does not put an accent on the reliable simulation of the HF noise. Notice that a procedure has been developed based on this standard by the General Motors Co for the bench tests intended to estimate NVH characteristics in brakes [80].

## *6.3.3 Experimental Equipment for Vibroacoustic Analysis*

Along with the traditional methods of noise and vibration measurements in brakes by the contact detectors (accelerometers) and measurement microphones, the informative contactless methods have become more popular recently. These are the Doppler laser vibrometry, electronic pulse speckle interferometry and acoustic holography. They share common drawbacks like intricacy and high cost of the equipment that restricts their broader application (Fig. 6.9).

The laser Doppler vibrometers (LDV) estimate vibration characteristics, the operating mode shape and perform modal analysis of the brake elements [81]. The

**Fig. 6.9** Brake noise studies by a full-scale dead chamber testing [86]

LDV operate at high resolution power of vibration velocity (till 0.02 μm/s) within a wide frequency range (0.05 Hz–22 kHz), shows the on-line fields of mechanical vibrations of the objects, and simulate their dynamic behavior. The experimental data may be used for computing sonic radiation intensity and simulating vibration in brakes. A procedure for determining acoustic activity of the brake systems based on the scanning LDV is discussed in work [89]. It presumes that eigenfreuqencies and mode shapes excited in a stationary brake system by an external vibration source are equivalent to the eigenfreuqencies and mode shapes generated by a brake generating squeal during operation. This equivalence forms advantages of above procedure. Firstly, there is no use in reproducing artificial squeal, which is hard to do in the lab conditions. Secondly, it furnishes a possibility to evaluate vibration on a stationary brake disc, by avoiding difficulties of vibration measurement on a rotating surface. Thirdly, the measurement results are independent of the brake torque variations since the vibration effect used to excite a stationary brake creates interaction forces between the lining and the disc varying at a frequency similar to that of the working brake system [90] (Fig. 6.10).

It often turns so that it is problematic to use LDV in vibration studies because of nonstationary behavior of the braking process. What is more, it is important for the researchers to evaluate not only the normal vibration components of the brake disc, but also to define its mode shapes and analyze its longitudinal components. There still remains a problem unsolved on the effect of tangential vibrations (due to instability of the braking forces) upon generation of transverse flexural vibrations known to be the cause of squeal. To make a spatial visualization of disc vibrations,

**Fig. 6.10** A general view of a testbed for SAE J2521 tests [87]

the authors of [91] have developed 3D scanning LDV systems to obtain mode shape images of the brake disc in different projections (Fig. 6.11).

The electronic pulse speckle-interferometry (EPSI) is a modern technique for contactless measurements of different objects in the whole deformation field scale under various loading conditions. The limitations of a unidimensional analysis (out of plane deformation) are avoided by determination of a complex 3D deformation vector using a 3D EPSI system. This system is capable of finding both transverse and longitudinal vibration-induced deformations of the brake disc. The EPSI performs measurements as follows [92]. The object under study is illuminated by short nanosecond laser light pulses. An optical signal from three different directions is simultaneously recorded by three cameras (Fig. 6.12). The EPSI results are reproduced as a spatial deformation field. The EPSI eliminates disadvantages of the known holographic methods [93, 94] and enables to study high-velocity processes [95–97].

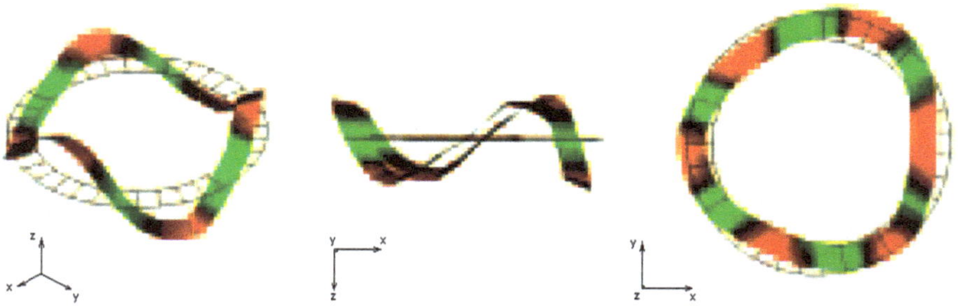


**Fig. 6.11** Brake disc mode shapes obtained by a 3D scanning LDV system [91]

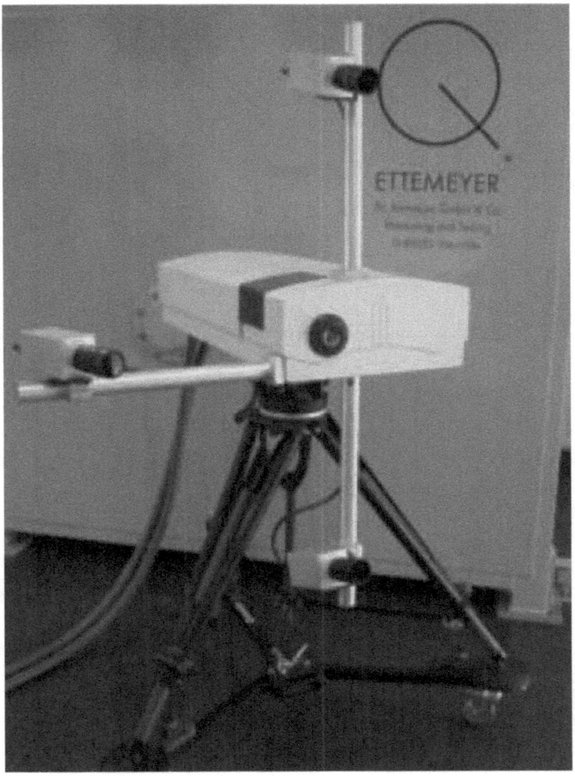


**Fig. 6.12** 3D EPSI system [92]

A 3D EPSI system operation for analyzing vibration of brake systems has been described elsewhere [92]. The data were evaluated in conditions of a bench test and during car movement. In the laboratory conditions the brake disc was excited by an electrodynamic vibrating rig.

The images in three different directions of sensor vision subjected to correction before calculations of the transverse and longitudinal deformations of the object are taken as the initial data. The corrected phase images in three directions are used to calculate longitudinal $V_x$ and $V_y$, and transverse $V_z$ deformation components of the disc under loading. Such a combination of different components in one chart makes the picture of vibration more vivid (Fig. 6.13).

Application of the method of nonstationary spatial transformations of the sound fields (nonstationary STSF) in the analysis of the high-frequency brake noise has been described in work [98]. Thus obtained animated contour map reflects the processes of the sound field formation and spreading in interrelation with disc vibrations in time. It is practically impossible to analyze sound fields at 16,000 shots/s frequency with high enough spatial resolution by any other existing methods, while STSF yields a detailed description of the sound field variations during braking with temporal resolution till a single oscillation time of the brake disc generating squeal at up to 4 kHz frequency.

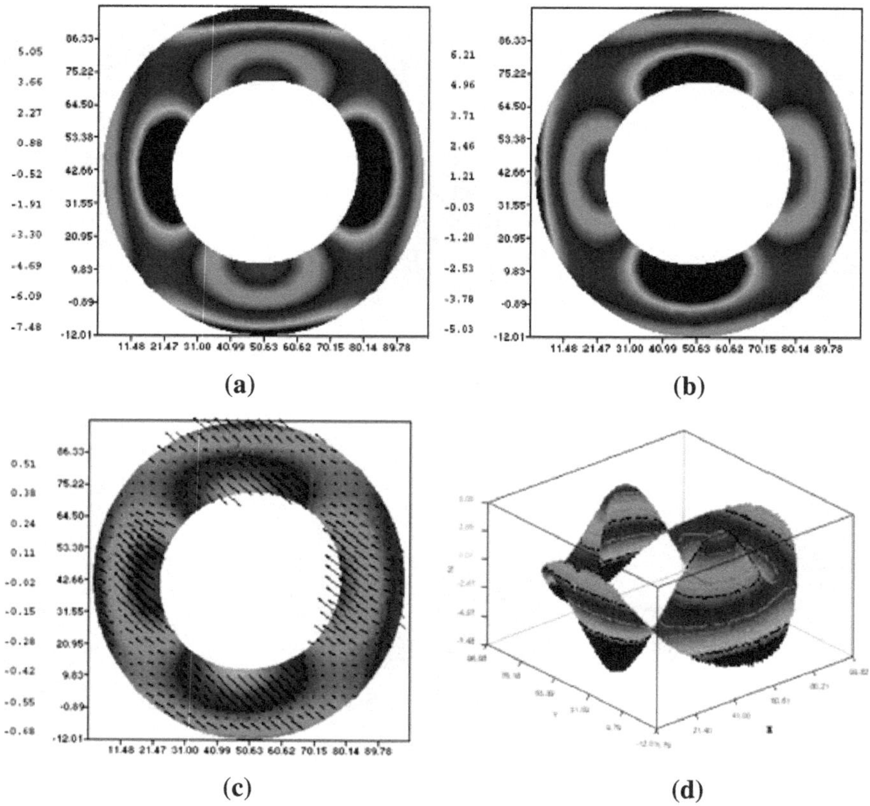


**Fig. 6.13** Noise and vibration measurement results of a brake disc ($f = 5{,}046$ Hz), obtained by the EPSI method in planes $V_x$ (**a**), $V_y$ (**b**), $V_z$ (**c**), 3D image (**d**)

The main drawback of this method is a limited spatial resolution since the oscillations exponentially damping with distance from the wave source can not be reconstructed fully. The instrumental resolution equals to 4–5 cm, which corresponds to the grid pitch. So far, the upper boundary of the frequency range under study is limited to the 3.2 kHz frequency.

## 6.4 Low-Frequency Forced Vibration

### *6.4.1 Forced Vibration Mechanism*

Forced vibrations or so-called hot judder of the automotive brakes are characterized by a directly proportional frequency dependence on rotation velocity of the wheels, and consequently, on speed of the vehicle. This type of vibration influences adversely driver's comfort perception, and due to its unexpectedness his response to traffic situations. Another adverse feature of the hot judder is cracking of the metallic brake disc due to a cyclic behavior of the mechanical and thermal loads. Friction-excited heat generation in the brakes and transmissions of the mobile vehicles are the cause of thermal warping of the contacting parts and variations in pressure distribution over the friction contact. High enough sliding velocities in conditions of nonstationary heating cause fluctuations in thermoelastic contact characteristics and may initiate thermoelastic instability processes. As a consequence, there appears instability in the low-frequency sliding velocity, or so-called friction torque variations (FTV). From the other hand, uneven wear of the friction pairs induced by thermoelastic phenomena or the metal disc run-out may also lead to FTV at brake application. The FTV are propagating from the source on the brake tribopair via suspension and body elements, and is perceived by the driver and passengers in the form of a local chatter of the steering wheel, brake pedal, judder of the interior parts, and a low-frequency structural hum.

The lion share of literature on the brake noise deals with the problems of the HF vibrations like squeal, and the corresponding mathematical methods of simulation and analysis. Much less attention is paid to investigations of the LF vibroacoustic phenomena, including hot judder and groan. More thorough attention is being paid to these effects in the automobile industry [68] and railroad transport nowadays [99].

The frequency of forced vibration depends in a general case on the wheel rotational speed. For instance, the frequencies with a doubled rotation frequency per second are called the second-order ones. We differentiate between two groups of vibrations according to their order [100]:

1. The low-order vibration. The order of this type of vibration makes up 1–5. Usually, some initially apparent deviations from the ideal geometry turn to be the reason of the low-order vibration or so-called "cold judder". Like in the case with inhomgeneity of thermophysical properties of the material, the result of geometrical imperfection is nonuniformity of the contact pressure and the temperature field generation, especially typical for the lasting and frequently repeated braking [101]. This is why, the initial cold judder may intensify and transform into the hot judder if braking prolongs.
2. Superposition of the low-order vibration due to geometrical deviations and/or friction-induced self-vibrations with the higher-order resonant vibrations. An example of superposition of forced vibrations arising from geometrical imperfection of the disc with resonant components is shown in Fig. 6.14. The figure presents a 3D FFT spectrum of a brake system vibration in the course of slow down. The amplitudes of the forced higher-order vibrations are not as a rule high. Nevertheless, these vibrations are gaining force during a long-lasting low-intensity braking. With increasing time of braking the temperature and pressure fields are gradually localizing. The hot bands appearing round the block or disc circumference acquire the form of repeated hot spots. The dominating order of the resultant vibration coincides as a rule with the actual number of hot spots [102, 103]. As for the low-order vibration, uneven heating brings about the short-term disc thickness variation (DTV) and deformation. Besides, high enough local

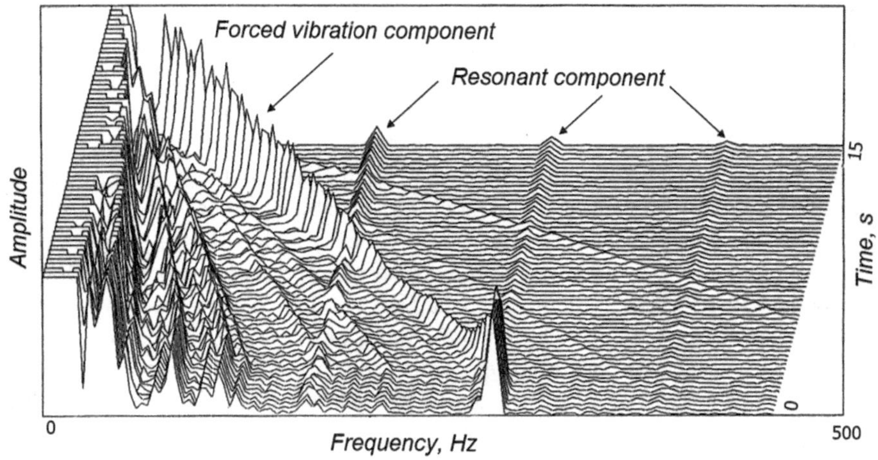


**Fig. 6.14** Superposition of forced vibrations with higher order resonant components of nonstationary friction

temperatures lead to irreversible changes in the friction and wear characteristics of the material on the corresponding surface areas and in the bulk. The upper boundary of the frequency range of the forced vibration is restricted by the maximal speed of the vehicle, as well as the wheel radius and vibration order.

There exists a number of factors able to initiate instability in the friction torque and pressure in the brake drive leading to forced vibrations. To these we relate the initial geometrical imperfection of the friction surface (nonflatness), nonuniformity of wear and transfer film formation on the tribopair, uneven heating of the disc and pressure distribution, inhomogeneous level of frictional characteristics and external forces (Fig. 6.15). Named phenomena similarly to the causes of their formation are not generally independent.

*Geometrical imperfection of friction surfaces.* From the viewpoint of geometry, the strongest effect on brake judder is exerted by the DTV and disc wobble. The amplitude of beating can be found from the production and installation tolerances of the disc, clearances in the bearings and disc deformation by braking. The DTV values may reach 15 μm and bring about significant vibration. In this connection most manufacturers keep to the fixed tolerances on the initial beating within 6–10 μm [104, 105].

Along with geometrical irregularities of statistical (preserved) origin, there exist dynamic (reversible) variations in geometry of the disc (thermal DTV, undulation, tapering, etc.). During operation the DTV is asymptotically growing due to wear from initial beating till the maximal value. The elevated initial beating may lead to a faster DTV growth.

The DTV may arise from a nonuniform corrosion of the disc in the case of the parking brake application, since the shoes are protecting a part of its surface.

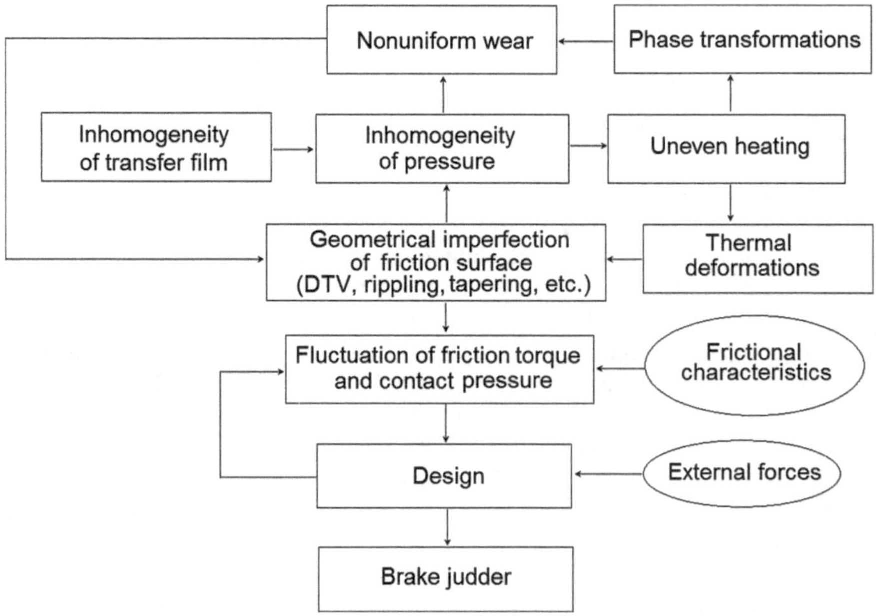


**Fig. 6.15** The factors leading to brake judder

Besides, beating is also influenced by the external forces appearing from a misbalance and interactions between the wheel and the road carpet.

In a general case, geometrical imperfections of the rubbing surface are in many ways connected with the frictional heating and wear processes. As the experimental investigations [104, 105] have proved, the major contribution into the brake judder is made by a short-term DTV.

The DTV are induced by numerous factors, such as:

(a) the initial DTV appearing at manufacture or mounting;
(b) the wear and processes of cleaning that intensify the DTV;
(c) both surface and volume properties of materials are always to some extent inhomogeneous [106] because of the phase transitions due to local overheating of the disc. In this case the DTV remain after the disc cooling;
(d) a temporary increase of thermal DTV takes place at each brake application because of the heating inhomgeneity, localization of contact regions and pressure as a result of thermoelasticity [101, 104]. Thermal expansion due to a local temperature difference by 200–300 °C leads directly to the DTV equal to $\approx$10 μm. So, the DTV increase with persistent braking, especially when the rigid shoes are used;
(e) variability of the transfer film thickness (islet character) contributes also to the DTV till a few micrometers;
(f) corrosion inhomgeneity and transfer of the heated friction material on the disc surface.

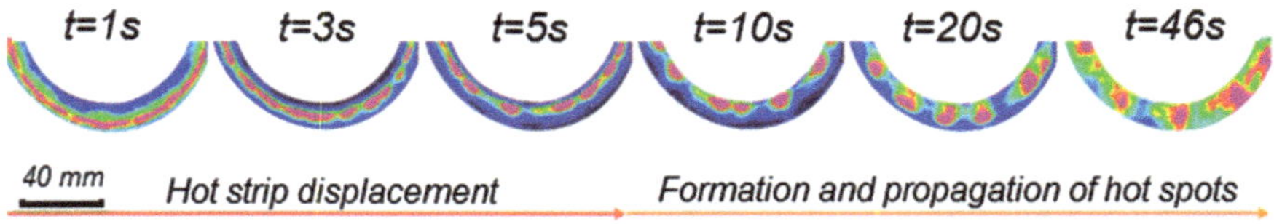


**Fig. 6.16** Hot spot formation on the friction contact under prolonged high-velocity braking [99]

*Uneven heating*. The disc brake sliding velocity and generated thereby heat are increasing with enlarging disc radius. Consequently, the conditions of uneven heat generation (temperature fields and pressure are localizing in the form of strips close to the outer radius) occur even if the plane-parallel conditions of the tribopair and homogeneous distribution of the friction coefficient are ideal [67]. These heated strips tend to transform into the "hot spots" with time of braking [107, 108]. The size of the hot spots exceeds much that of the roughness value but cedes the friction contact area width in the way like Fig. 6.16 shows.

The maximal temperature on the hot spots is reached as the rigidity of the brake shoes is increased [109]. The measurements have shown that the local temperature on the hot spots can reach as much as 700–800 °C, while the temperature difference on the friction surface of the disc is about 300–600 °C [66, 67, 110]. The hot spots are usually distributed randomly. Table 6.1 presents the comparative characteristics of the main types of hot spots according to the classification discussed in [107].

The intensive heating of the disc in the hot spot region leads to its uneven thermal expansion or a heat-induced DTV. This process can be unstable, so we may relate it to thermoelastic instability (TEI). As the time of braking extends, the TEI adds to the localized character of the pressure and temperature fields. In a limiting case, especially when the high-order oscillations are excited (6–20), the disc undergoes cracking.

The investigations conducted by the authors of [111] have supported the assumption that less dimensions of the disc (disc and pad thickness, friction path diameter) intensify the tendency to the hot spot formation. In addition, elevated sliding velocity and resulted friction energy are also promoting formation of the hot spots.

The main reason of forced vibrations excited by a lasting or repeated braking is rather the temperature gradient than the total thermal load elevation on the brake [112]. The temperature gradients cause instantaneous DTV due to the inhomogeneous thermal expansion of the disc material [66]. In the case a rigid brake block

**Table 6.1** Comparative characteristics of hot spots

| Hot spot type | Maximal size of hot spot, μm | Friction surface temperature, °C | Lifetime of spot, c |
|---|---|---|---|
| On the roughness | <1 | 1,000–1,200 | $<10^{-3}$ |
| Central | 5–20 | 750–1,200 | 0.5–20 |
| Transverse | 20–100 | 100–700 | >10 |
| Local | 50–100 | 10–100 | <10 |

both the DTV, and FTV increase show a relatively linear behavior at prolonged braking [101], especially over the outer radius. That is why the time of braking and brake application frequency during a cyclic slow down are considered as the critical factors in the appearance of the forced LF vibrations [112]. The prolonged braking and the corresponding brake torque reduction promote the formation of the hot spots and heat-induced DTV [109]. So, the prolonged slow-down and reduced friction torque elevate the temperature and pressure gradient as compared to a short-term 3–4 s lasting heavy-loaded braking [101, 109, 113].

The hot spots are formed faster when the sliding velocity increases with growing energy load on the brake [104, 111]. The localization processes and the correlated DTV are running at ≈100 km/h velocities [104]. This is because to initiate TEI some minimal critical velocity is to be reached. In this connection, the standard tests of frictional materials are conducted at >100 km/h speeds.

*Phase transformations*. The contact surfaces of the rubbing bodies are exposed to the intensified heat loading in the hot spot regions, which gives rise to phase transformations in the materials. Gray iron experiences the strongest phase transformations under nonstationary friction. The areas with irreversible martensitic structures of 650–800 HV hardness can be formed in the hot spot regions when the temperature is above 740 °C, and the cooling is not complete (below 300 °C at heating rate above 500 °C/s). Figure 6.17 illustrates the dependence of microhardness on a hot spot area of the material through depth from the friction surface [114].

The process of transformation into the martensite depends on carbon content in the matrix, and to a lesser degree on concentration of other alloying dopes. The volume in the heated region may expand by up to 40 % as a result of martensite transformation. Named changes are preserved when the temperature drops, consequently, they continue to influence geometry of the rubbing bodies and tribological properties of the tribopair. Martensite formations at the macrolevel may generate high local stresses that activate deformation processes and cracking of the cooled surface in the heat-generation cycles that follow. As a result, the residual strains in the disc may induce cold judder not connected directly with further heat

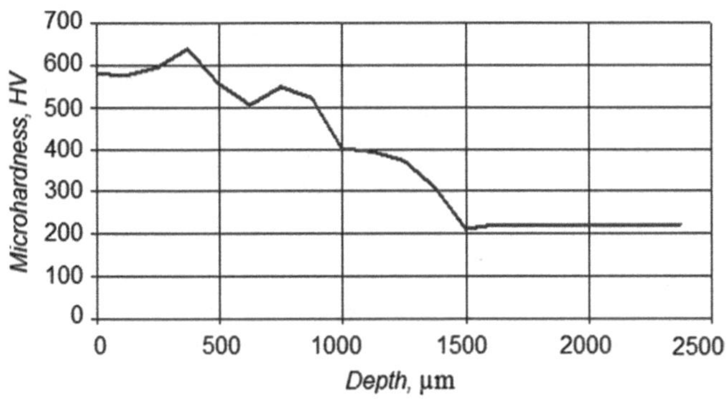


**Fig. 6.17** Distribution of microhardness through disc depth in the *blue* spot region resulted from the local microstructural phase transformations [114]

generation in the friction pair. Formation of metallic carbide disks on the surface presents more serious, although less probable problem [114, 115].

*Inhomogeneity of contact pressure.* Experimental observations have proved that braking induced heat generation and, consequently, contact pressure distribution in the radial plane are far from being homogeneous. They actually display a discontinuous behavior in the form of small areas remaining invariable at single braking (Fig. 6.16). The position of hot spots may vary in the repeated braking events [111]. Further on, pressure localization on the friction contact may be influenced by the increasing thermal deformation of the brake disc due to inhomgeneity of the *heat flows.*

*Thermal* deformations. Thermal deformations of the brake disc are subdivided into the following types:

1. *Undulation.* This is most probable type of thermal deformations that exerts an essential influence on the actual contact area. The wavelength of the disc is conditioned by a number of factors, on the first place, by a stationary temperature gradient between the friction surface and the hub.
2. *Conicity.* In the case of this type of heat-induced deformation, the friction surface deviation towards the hub may reach 200 μm and result in beating [116]. As it was proved in [66], conicity-induced beating depends very much on the disc design.
3. *Uneven thermal expansion.* The temperature difference of about 250 °C, which is not a rare case, results in the DTV with a 10 μm [66].
4. References [66, 117]. Momentary distortions in disc geometry may bring about forced vibrations even if the hot spots are being formed during braking.
   Some of frequently observed forms of thermal deformations of the brake disc are illustrated in Fig. 6.18.

Above-described mechanisms of thermal deformations are initiating, as a rule, a short-term hot judder. When the temperature gradients are high enough and prolonged, thermal strains may cause irreversible changes in the disc shape and a continuous cold judder. It is shown in [118, 119] that the cause of persistent thermal deformations may be also the regions of residual restoring stresses occurred at manufacture of the disc, especially if flatness has been reached by grinding.

*Inhomogeneous wear.* Wear of the brake disc in conditions of a disengaged brake motion is sometimes called "cold erosion". It is attributed to the DTV and may lead to a judder. If the disc runs-out at a released brake, the lining might

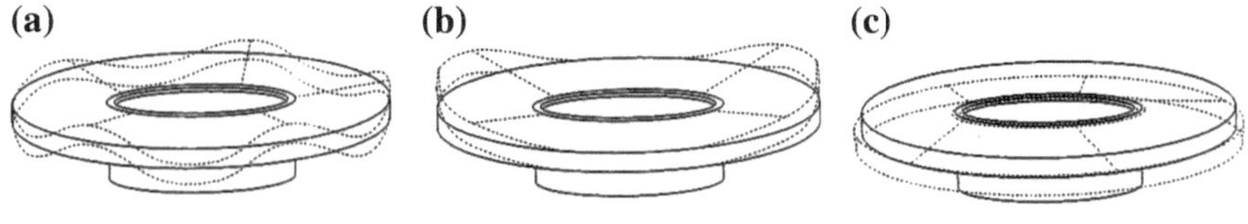

**Fig. 6.18** The main forms of thermal deformations of the brake disc: **a**, **b** undulating, **c** conicity [114]

periodically touch slightly some portion (not the whole) of the disc thus initiating its inhomogeneous wear.

In the case the DTV is caused by the periodical contact of the lining, two competing processes may arise, i.e., the formation and elimination. The domination of one or another process depends upon the friction material used. The friction materials provoking the DTV when the brake is released are able to eliminate the DTV via application of the brake, initiating thereby a stronger wear of the disc. The intensified wear of the lining at elevated temperatures and pressures may, in particular, weaken the danger of the hot spot formation thanks to a more uniform contact pressure distribution over the friction surface [67, 120]. However, the intensified wear is accompanied by accelerated shear of the hot spots [67, 109, 120], which is the reason of thermal fatigue of the disc [113].

A careful driver would notice the appearance of judder by far earlier than the one fond of spin, because the gradual application of the brake at a low pressure of the lining hampers clearing of the disc (DTV elimination). The discs that trigger pressure variations in the brakes (BPV) and the accompanying forced vibration can be in some cases corrected by a few emergency braking acts. However, the variations in characteristics of the materials during operation are contributing more in the DTV than the style of driving [121].

The wear, especially that of the lining, increases significantly if the braking lasts long. The elevated wear equalizes the temperature field and reduces the maximal temperature value [120]. Therefore, the role of wear turns to be positive due to removal of the hot spots and bands that together with the thermal fatigue of the disc may lead to cracking. The results of simulation [67, 120] and experiments [122] have proved that the intensified high-temperature wear of the friction lining prevents contact pressure from localizing on the friction surface and may decrease generation of the hot spots.

*Inhomgeneity of transfer films*. The third-body layers or the transfer films present the layers about a few micrometers thick consisting of wear debris formed by the rubbing bodies [99, 122]. The metallic particles resulted from the gray iron disc wear are oxidation in the atmospheric oxygen and deposit on the friction surface in the form of a grayish-black film. The properties of the films on the friction surface and their thickness uniformity define the friction characteristics of the brakes and the FTV level.

When a heated brake is applied in a stationary vehicle, the lining might stick to the disc, while at $\sim$500 °C temperatures the molten fictional material sinters onto the disc.

*Frictional characteristics and the friction coefficient value*. The dependence of the friction coefficient upon sliding velocity is traditionally considered to be a source of all types of vibrations in the brakes, including the forced ones. Jacobson [68, 123] has analytically confirmed that forced vibrations in brakes occur even if the friction coefficient is invariable. Forced vibrations may arise in brakes independently of the tribopair characteristics, although the ascending dependence of the friction coefficient on the braking pressure may intensify vibration. To reach a desired brake torque when a friction material with a lesser friction coefficient is employed, one should use a higher pressure. In this case, the contact pressure and temperatures are distributed more uniformly [113]. Consequently, the reduction of the friction coefficient exerts a positive effect on the problem of forced vibrations and decreases the propensity to cracking. Besides, the relationship between the DTV and BTV is a function of the absolute value of the friction forces. This is why the FTV amplitude is proportional to the friction coefficient [104], while DTV is independent of friction.

*External forces*. Different oscillations excited by a disbalance or variations in the forces initiated by the tires due to the elastic properties of the hub and bearing elements are transmitted onto the brake disc. The external effects generate the vibration multiple to the instantaneous frequency of the disc and similar to the one caused by geometrical imperfections of the disc. The external forces are able, therefore, to influence forced vibrations. The structure "wheel—hub—bearing" and its stiffness are governing the amplitude of disc deviations [104]. So, it will be incorrect to consider forced vibrations of the brake caused by the tires or disbalance as the "brake vibration" since it implies that their source is the brake itself.

*Design factors*. The structural elements of the suspension are perceiving large oscillation amplitudes in vertical directions along with the longitudinal ones caused by the brake vibration [123, 124]. The structural peculiarities of the front suspension, and its lengthwise stiffness in the first place, is dominating in the vibration transmission from the brake onto the car body, as well as its perception by the driver and the passengers [124]. The investigation results discussed in [105] prove that rigidity of the bushes (silent blocks) of the lower arm is a critical parameter since it defines the frequencies of self-excited vibrations of the suspension longitudinally.

A detailed study of the suspension design effect discussed in [125] has singled out the factors that may spur forced vibrations in brakes:

(a) bushes of the tie bar contributing to radial and longitudinal stiffness;
(b) the lower arm bushes responsible for longitudinal stiffness;
(c) the antiroll bar proving vertical stiffness.

The frequency of self-induced oscillations of the suspension longitudinally is dependent on stiffness of the lower arm bushes. By rising the frequency of self-oscillations it is possible to decide the problem of brake vibrations either by increasing stiffness of the rubber bushes or reducing modal masses and inertia of the suspension [68].

Nevertheless, the modern suspensions design with a lowered longitudinal stiffness is considered to be advantageous because of a radial-type of the tires [124]. A decreased longitudinal stiffness of the suspension is needed to dissipate longitudinal oscillations from a rigid tire belt.

It is to be noted that the self-oscillation frequencies and damping should be measured only in the position when the brake is engaged. The experiments have shown that the eigenfrequency increases considerably when the brake is released (18 Hz in contrast to 13.8 with the engaged brake), while the equivalent coefficient of viscous damping reduces (0.07 in contrast to 0.08 with engaged brake) [123].

### 6.4.2 Investigation Approaches to Forced Vibrations in Brakes

To study forced vibrations in brakes one should know the prehistory of braking within different time scales. We acknowledge here three time scales:

(a) time of a single wheel revolution;
(b) time of braking or the time between two sequential braking events;
(c) lifespan of brake components.

The designers usually accept a long-term scale in the brake construction, intending to account for the changes occurring within their service life. Of no less importance is to study the processes accompanying braking. To these we relate localization of the contact areas, temperature and pressure gradients, formation of the hot bands. As the investigations show, the DTV is a dynamic characteristic able to vary much during the braking cycle [101, 104].

The processes occurring within a single wheel revolution are neglected as a rule due to intricacy of their numeric analysis and processing by the traditional methods. The sinusoidal changes in the brake torque per single revolution of the wheel takes into account the amplitude functions described in [112].

The peculiarities of a viewpoint on some problem and its analysis we shall further call an approach to the problem, meaning a theoretical abstraction or representation of the world that includes its model and limits.

Any approach to a problem highlights the physical phenomena to be studied, the time and spatial scales to be used. The methods employed for analytic and experimental works as well as the type of solution depend upon the chosen approach to the problem.

The analysis of literature devoted to forced vibrations in brakes has shown that the approaches used can be definitely divided into two groups, namely, the casual and investigatory approaches. The investigatory approach estimates the vibration sources proceeding from the amplitude and frequency (order) parameters affected by the FTV and/or BPV. In contrast, the casual approach studies such physical phenomena as wear and heating.

There are several types of casual approaches [68]:

1. *The systematic approach* studies the initiation of forced vibrations as a result of a complex effect of the vehicle structure as a whole or the suspension with the FTO/BPV. In this case, the braking frequency is considered constant or changeable in a parametric way. This approach is mostly used in the experimental research [126, 127]. It is also used to analyze quasi-static processes. It is interesting to know that the method of a multiparticle system analysis has been used by Kim et al. [128] to study the dynamics of a McFerson suspension on a model with 12 degrees of freedom excited by the first-order forced vibrations of the brake.

**Table 6.2** The approaches and corresponding methods

| Approach | Methods |
|---|---|
| Casual approach | Finite-element method, method of finite differences, bench and labtests, measurements of temperature fields, X-ray methods |
| Systematic approach | Multiparticle system analysis, Fourier method, modal analysis, Taguchi method, ride tests, spectral analysis |
| Swept frequency approach | The analysis in time region, ride tests at braking, construction of 3D spectra, ordinal analysis. ABC and capacitive transducers |
| The approach of subjective estimates | Subjective estimate of drivers and passengers, braking tests |

2. *The swept frequency approach* studies the forced vibrations excited by the braking proper along with the FTV/BPV-induced frequency variations. The dynamic characteristics of a vehicle design can be presented by the actual critical velocity resonance [112].
3. *The approach of subjective estimates* studies the perception of forced vibrations by a test driver or a common man in a test car. Along with determination of the vibration level and frequency, the subjective estimate can be used as an alternative for finding weight, height and location of the investigator in a car.

Table 6.2 presents most applicable approaches with respective analytic and experimental methods [68]. For instance, the finite-element method is a key tool in FTV examination caused, in particular, by the TEI. The casual approach does not consider the initial phenomena leading to the FTV, whereas the result is simulated by a sinusoidal driving force that excites the braking system and the conjugated elements. This assists in analyzing a multiparticle system (such software packages ADAMS, DADS and other) as a key and highly important one for the Lf vibrations (below 50 Hz).

The experimental studies of forced vibrations in brakes are usually conducted in the form of the bench and ride tests. The main advantages of the bench tests [104] are the accurate reproducibility of the test conditions, high-sensitive measurement devices, moderate financial and time efforts [109, 129]. The disadvantages are in the difficulty to consider all other elements of the vehicle that contribute to transmission of the BTV (the tires, suspension, steering gear) [104]. To the drawbacks of the ride (road) tests we refer the difficulty to control the BTV/DTV during braking. Nevertheless, this problem can be solved by measuring the BTV/DTV instead of governing these processes artificially by maintaining the required speed, temperature, pressure and so on [102, 112].

### 6.4.3 Variations in the Brake Torque and Contact Pressure

The sources of BTV and BPV are commonly localized. We should, however, bear in mind that these sources are brought about by interactions between different in nature phenomena (see Sect. 6.4.1). Beating and the uneven disc thickness,

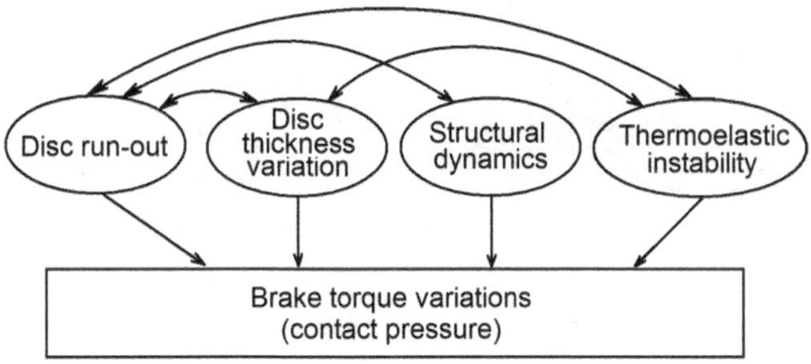


**Fig. 6.19** Forced vibration causes in disc brakes

i.e., geometrical imperfections, cause variability of the normal forces and, as a consequence, lead to the BTV and BPV (Fig. 6.19). In addition, the effect of thermoelastic instability contributes to the onset of the hot spots and sometimes to persistent areas with the altered physico-mechanical characteristics, which may intensify vibration.

Of no less importance is contribution of the external factors into the elasto-dynamic deformation of the disc that are induced by the tire interaction with the road carpet and the disbalance phenomena.

The traditional design methods of the disc brakes make possible to define the temperature and pressure distribution, the maximal temperatures, pressures and stresses, which turn to be much less than their actual values [67].

One of the design methods for thermal deformation calculations in the brake elements during deceleration is discussed in [130]. This method is based on the analytic calculations of the brake system temperature with account of its cooling. The results are further used as the initial data for the numerical solution of the unbound problem on the thermal stress–strain state.

The investigations of the friction pairs by solving the unbound thermoelastic problem using the finite element method have shown that hot judder is most probable under a relatively low contact pressure and high overall power of the brake [66].

Figure 6.20 illustrates the calculation results of the stress–strain state of a brake disc based on the unbound thermoelastic problem solution. It is seen from Fig. 6.20a that the mechanical stresses induced by a cycle of braking events are concentrated between the mean and outer friction radii of the brake disc. Thermal deformation may take the form of undulation (Fig. 6.20b) or nicity (Fig. 6.20c).

For investigations of forced vibrations in brakes it is important to solve the bound thermoelastic problems in which the mechanical and heat loads are considered in combination. The available commercial software packages of the finite-element analysis propose separate solutions of the deformation and thermal problems. The authors of [131] have presented a solution of a 2D (axisymmetric) bound thermoelastic problem that simulates the formation of hot bands on the brake disc. The main idea of the method consists in a sequential switching from the mechanical to thermal problems, each being solved by a commercial software of the

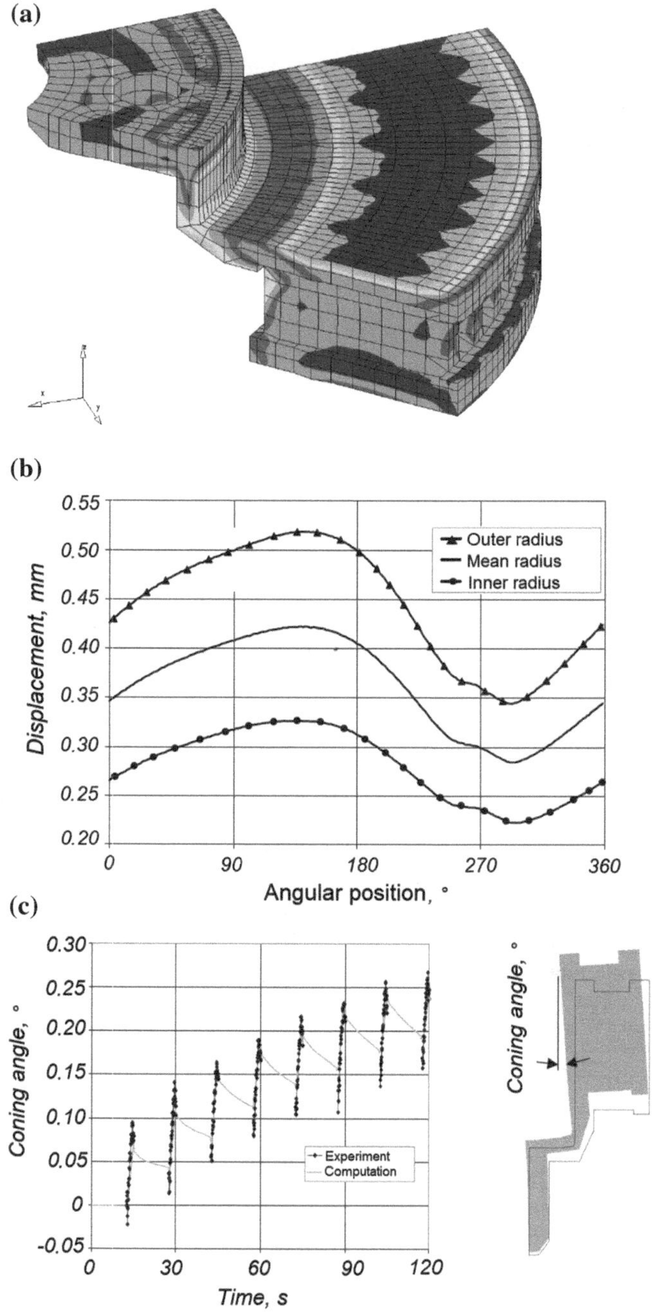


**Fig. 6.20** The computation results of mechanical deformation of a brake disc by the finite-element method: **a** Moses stress outline using the finite-element model after 8th braking cycle; **b** side deformation; **c** conical deformation value versus braking time [130]

finite-element analysis (ABAQUS, LS-Dyna, Ansys or other). The finite-element model considers the time dependence of a certain region. An analogous concept is discussed in [101]. To study the effect of structural parameters of the block on the heat-induced cracking, simulation of the thermoelastic warping and DTV, the authors have used both a 2D (axisymmetric) and 3D (to simulate warping) models. The solution of the bound thermoelastic problem by the finite-element method and the mixed-type elements involves wear simulation. It has shown that at a single braking the hot spots might slowly shift as the wear on their nascent regions increases [67].

Simulation of TEI processes in a brake system with account of the actual braking conditions toughens the requirements to the hardware thus complicating the computation, processing and data storage [101]. This problem is especially acute in simulation of a slow retardation (so-called stiff problem) in the mode of a long-run braking. Notice, that just this type of braking contributes a lion portion into the formation of hot spots. The problem of raising the calculation efficiency can be solved by one of the next approaches:

1. It is possible to disregard the circular measurements in the instantaneous state of the surface and disc geometry. It should be noted that circular measurements play an important role in the low-frequency vibration of brakes.
2. The use of a special type of contact elements enabling to reach a needed friction force and heat on the contact surface [67, 129].
3. Simultaneous solution of the problems of interrelated areas as a most accurate and efficient one [120]. Since it is based on the Newtonian method developed in the 1990s, its algorithms are not included into all software packages of the finite-element simulation.
4. The use of a 3D hybrid method combining the fast Fourier transform (FFT) with the finite-element methods. This method has been developed by Floquet and Dubourg [132]. The FFT (for spatial variables) makes possible to decrease the problem scale. The variables are transformed into the discrete frequency parameters and the corresponding frequency derivatives are excluded. The method is applied to geometrically periodic solids devoid of the axial symmetry like, e.g., the vented brake discs.
5. The direct use of a small parameter method (perturbation) in the finite-element analysis. The main idea consists not in solving the nonstationary problem but in considering only the conditions at which slight variations of the temperature field promote their exponential growth in time. This method is used for stationary processes where the contact area is constant in time. In reality, the contact area varies essentially with the braking time.
6. The use of a brake block model that involves viscoelastic elements [111].

The methods of computation hydrodynamics (CHD) are often used to study the vented discs, in particular, aluminum discs for which the air flow is very important. These methods are used by the brake disc manufacturers to widen the air flow [133, 134]. The marketing has shown that some companies recognize the vistas in achieving more accurate design data on the heat loads in brakes by combining the finite-element analysis and the CHD method. It should be noted, however, that the related calculations are rather laborious.

The approximate approaches may be used as an alternative to above-mentioned ones that estimate the convective cooling. The designs of the ventilated brake discs employ one of the next two types of convective cooling: by a transverse air flow over the disc surface or by an air flow through the ventilating channels [67]. Simulation methods of thermal processes in the brake systems have for long used the relations that estimate the convective heat exchange factor in the brake disc. Today we commonly use two dependencies to calculate the air flow in the ventilated disc, namely the Sisson's and Limpert ones [135]. As an alternative it is possible to approximate the convective heat exchange factor by the Nasselt module [119] followed by its correction with account of the empirical parameters.

### 6.4.4 Simulation of Forced Vibrations

In spite of a broad discussion in scientific literature of the brake vibration effects in vehicles due to BTV and BPV, the interest to these phenomena is still acute. The results presented in [121, 136], and many other works agree in the fact that the maximal vibration levels appear in the brakes at definite speeds. However, it is extremely hard to calculate and simulate these processes and the results obtained are unreliable. In this connection, the researchers have to use the subjective estimates for such studies with a 10-mark grading system. It turns so that the problems involving forced vibrations are being solved today rather by the trial-and-error method than by any systematic procedures.

There are numerous works in literature that analyze sensitivity, but they do not actually deal with a "pure" modeling. The pure modeling presumes the calculations of brake vibrations in the time region, including simulation of not only characteristics of the vibration sources (BTV and BPV) and their interrelation with the structural elements (resonances and their spreading) but of the very braking event too (frequency sweeping, braking conditions) as Fig. 6.21 shows.

It is considered inefficient to study forced vibrations in brakes by the direct integration of differential equations using common algorithms (like Runge–Kutt) developed for the non-stiff problems and devoid of the rational transformations of the variables and respective assumptions. The calculation time may be unjustifiably long (from a few hours till several days), while the result is often erroneous. The algorithms of the commercial software used in mechanics, e.g., ADAMS, turn to be rather slow and inefficient when a complete braking cycle is simulated. Therefore, the maximal vibration levels are often overstated within the limited time intervals of braking. Nevertheless, mentioned software packages work successfully in the quasi-static analysis of sensitivity.

*Sensitivity analysis.* The problems on forced vibrations solved by a systematic approach admit that not the whole vehicle is prone to vibrations, therefore, its sensitivity to the BTV and BPV is lower. For instance, Engel's system [126]

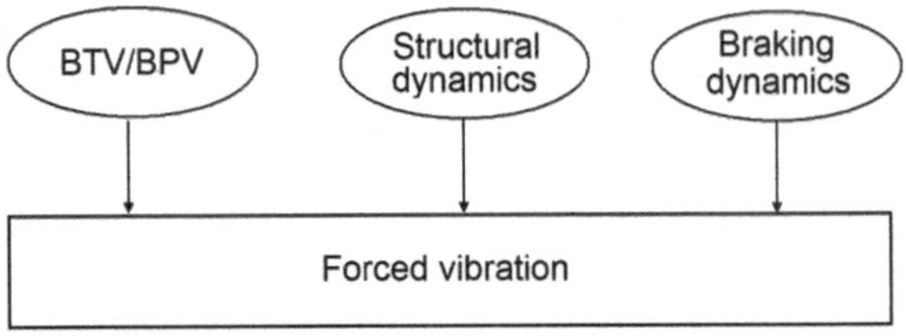


**Fig. 6.21** The analysis of sensitivity by a swept frequency method [68]

includes five different elements: a brake disc, a support and shoes, a tire and a wheel hub joint, spherical joints of the steering, and a steering control unit. Such a system is fit with a feedback between its elements.

The numerical simulation of a vehicle susceptibility to BTV has been described in work [128] by the multiparticle analysis on the example of above-named system. Unfortunately, we are not aware of the suspension model used for simulations. The analysis of this kind is employed in automotive industry but is little known to a broader audience. There is much information in literature devoted to Augsburg's model [104] incorporating the following elements.

(a) a support represented by two masses connected via a spring simulating lengthwise elasticity of the support;
(b) brake shoes—by a set of springs;
(c) a brake piston—by a mass;
(d) a hydraulic system—by the volume-accumulating elements.

The path over which vibration propagates from the source till its contact with the driver can be quantitatively studied on the base of simultaneous measurements of vibration accelerations in such points as the steering column, the steering tierod, a wheel, and etc. the sampled signals between separate points can be used to obtain the corresponding transfer functions [102, 127]. Above-described procedure can be used to find the forcing effects and transfer functions that characterize sensitivity of a car.

The amplitude functions [112] can be considered as the generalized transfer functions. They may be used to classify the braking events and generated by them vibrations. In contrast to the method of transfer functions, that of the amplitude functions can be used at a much higher retardation thanks to independence of the FFT. This method estimates eigenfrequencies more accurately since it takes into account the time lag of the maximal amplification caused by the finite retardation and inertia of the system. In the case the time lag is not accounted for, the eigenfrequencies may become systematically underestimated, especially when the retardation is high.

Investigations of the braking process within some frequency region have shown that the vibration frequency in the beginning of braking exceeds some limit that makes up 10–30 % of the critical speed. Braking is accompanied by a clear shear of the vibration amplitude maximum, which is especially evident in the systems with a poor damping. For the 1st order vibration studies it is desirable to use the methods of nonstationary processes (including swept frequency procedure) if damping is about 1 %, except for the case of infinitesimal retardations [68].

The braking process with a slow deceleration has been studied in work [137]. The authors have come to a conclusion that the peak of the vibration amplitude appearing under a concrete speed is a result of the drop of the kinetic characteristic of friction used in frequency calculations. To our regret, the authors did not consider the frequency sweep as a separate phenomenon.

*Simulation of forced vibrations.* In most of design and experimental studies of forced vibrations in brakes, braking is studied at a constant sliding velocity and, consequently, the constant rotation frequency of the wheel, the constant pressure and temperature. It seems reasonable for simplification of the measurements and more efficient achievement of data in the frequency region. However, vibration in the real braking processes is not a constant value.

> The increased level of BTV and BPV is observed when the speed either coincides or is close to the critical one. The vibration is perceived as soon as the brake is applied. The maximal vibration amplitude is reached at a certain speed. At deceleration, the vibration is preserved if the brake is continuously engaged. This is typical for the forced vibration generated by a constant frequency source and a sweeping frequency.

Figure 6.22 [123] illustrating the rotor–stator model with two degrees of freedom and a linear deceleration of the disc describes the main properties of the forced vibration in brakes. Prior to calculations, some nonoscillatory parts $\Phi$ and $\Phi_C$ in the angles of shift of the rotor (disc) and stator (support), correspondingly, and superposition vibrations $\varphi_D$ and $\varphi_C$ are isolated. The form of the amplitude function $E$ describes quite accurately the relative level of vibration $\varphi_C$ (second-time derivative of $\varphi_C$). All parameters of the rotor–stator model can be determined experimentally.

A full-scale model of a vehicle presented in work [138], has shown analogous vibration parameters, and is able to explain a number of other phenomena. However, this model is more complex and requires additional information, e.g., on wind velocity, cohesion factor to the road carpet and other.

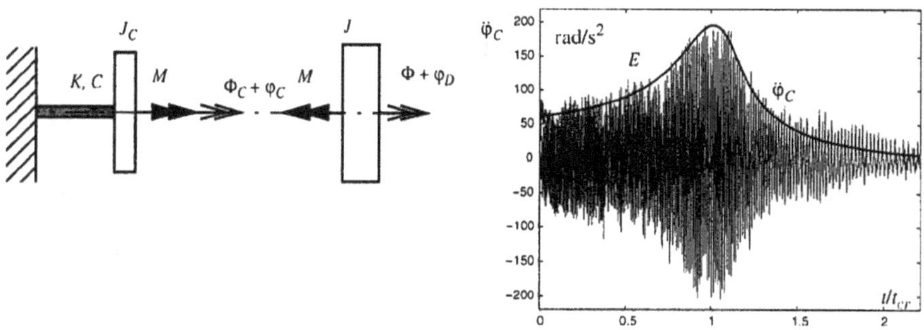


**Fig. 6.22** A model and simulation results compared to experimental data [123]

In designing passenger cars, stiffness of the front suspension should ensure the frequency of the longitudinal resonant vibrations in the range 10–20 Hz in order to make this resonance generate at a corresponding speed. In this case, the maximal vibration amplitudes for the first-order forced vibrations at a critical speed are observed within 60–140 km/h. For the second order forced vibrations the corresponding critical speeds (and accompanying resonances) will be within 30–70 km/h range. The resonant frequency for the car under study is about 14 Hz, and the critical speed corresponding to the first order is 95 km/h.

The low-frequency forced vibration in brakes has been insufficiently studied for the time being. A series of additional profound investigations are to be fulfilled in order to understand the urgent problems in this field. We should mention some challenging directions [68, 99, 139–142] that present a special interest in this research domain:

(a) simulation of the BTV initiation with account of wear and thermoelasticity (aggravation of wear, especially at high temperature and pressure reduces TEI);
(b) interrelation between the forced vibration and friction-excited self-vibrations;
(c) inhomgeneity of transfer films;
(d) TEI processes and their relation with warping;
(e) investigation of vibration modes of the suspension at braking;
(f) the BTV interrelation with simultaneous variations in shape (DTV, beating, equivalent radius) and frictional characteristics;
(g) calculations of the BTV level and order for preset design parameters and braking conditions;
(h) design of computing models for vibration values in cars at a given BTV level.

In a general case, the investigations of forced vibrations in brakes consists of two stages:

1. The analysis of the excitation sources (e.g., BTV, etc.) that includes a thermomechanical analysis of the brake system components by the finite element method or bench tests. Both computations and experiments at this stage are extremely time-consuming.
2. The analysis of the excitement source effect in a car. If the forcing effects are known (from calculations by the finite-element method or the test measurements) and preset in the form of time functions, one can analyze a system response to the excitement by above-described method of the amplitude functions.

The method of amplitude functions needs further elaboration, although it presents no difficulties for analyzing the multimass system model of the whole vehicle. Named models may be useful in finding the criteria for perfecting brake components or the suspension, or else.

### *6.4.5 The Methods of Forced Vibration Abatement in Brakes*

The general approaches to decreasing forced vibration in brakes were formulated, e.g., in [68] as follows:

(a) decreasing of BTV and/or BPV (casual approach);
(b) increment in the relative mass and inertia moment of the stator in contrast to the rotor;
(c) amelioration of damping;
(d) reduction of energy load in brakes;
(e) elevation of eigenfrequency in the brake system.

By decreasing mass of a car and its wheel radius it is possible to lower forced vibrations in a brake [68]. What is more, the reduction of the car mass brings down heat-induced DTV, the appearance of hot spots and the risk of thermal cracking in conditions of invariable heat capacity of the disc. The reduction of the stator mass (i.e., of the post and disc in the last turn) increases vibration because the DTV intensifies in this case.

The vented brake discs have gained popularity thanks to the light weight and improved conditions for the convective heat exchange. It is shown in [66] that variations in the vented brake disc design by changing location of the friction ring leads to a taper-like deformation only in the opposite direction and decreased till 100 μm magnitude. The back-vented design of the brake disc is widely applicable, although its air cooling is less efficient as compared to a through-vented design.

In spite of above advantages, the vented discs may induce brake vibrations by inhomogeneity of their thermal fields. The heat capacity of a disc in the case of a short-term emergency braking (less than a minute) from a high initial velocity is a more important factor. This is why, the solid brake discs with a larger mass turn to be colder than the vented ones [67, 119]. However, when the braking is prolonged, the temperature of the vented disc rises significantly [119].

In the solid (unvented) brake discs in which the design does not admit rearrangement of the friction ring, the potential tapering may be prevented by making special slits along the inner diameter of the ring in the form of incisions, or cooling grooves. This minimizes tapering for most of the solid and through-type vented discs (the internal hub-mating surface).

The approaches to abating forced vibrations are considered in detail from the standpoints of materials science in Chap. 7.

## 6.5 Low-Frequency Brake Noise (Groan)

The effect of groan generated by the car brakes is a frequent phenomenon that irritates both the driver and the passengers. The groan frequency range coincides with judder (forced vibrations) according to the classification set forth in Sect. 6.2.

Since these two types of brake noise differ in their manifestations (generated at a low speed and pressure) and physical mechanism of excitement (friction-induced self-vibrations of the first kind), groan is isolated into a separate class of brake noise.

### 6.5.1 *Experimental Investigations of Groan in Brakes*

The scientific papers dealing with experimental investigations of the brake groan pay first and foremost attention to determination of the friction surface characteristics (kinetic and static parameters of friction, in the first place) and their influence on the friction-excited self-vibrations. The research performed in this sphere has proved the possibility of improving vibroacoustic characteristics of tribojoints via optimizing their structure and composition of the friction materials used. The physical aspects of the friction-induced self-vibrations as the main cause of groan have been already discussed in Chap. 5. The problems concerning how to decrease the brake noise will be considered in Chap. 7 from the viewpoint of materials science.

The earlier experimental investigations [64, 143–145, 147], etc. were helpful in determining the dynamic characteristics of the brake systems but they are lacking information on the whole suspension design. The effect of design parameters, including suspension components, upon the processes of excitement and propagation of self-vibrations in the cabin, and their relation to groan generation have been studied in more recent works [70, 146].

The experimental investigations of the brake groan in a car fit with a McPherson suspension have been discussed in [70]. The test car was installed on a bench with a race drum. The meters were placed as shown in Fig. 6.23. The accelerometers were fastened on a strut from beneath the lower spring and on the support. The microphones were installed on the wheel arc and in the cabin. The rotation frequency of the race drum corresponded to a less than 1 km/h speed. The brake is able to quickly prevent the race drum from turning, whereupon it is gradually released generating a groan in this moment.

The experimental data obtained by different authors on the brake groan have proved that together with tangential vibrations of the brake support conjugated with other brake units it also experiences the vibrations in vertical direction [69, 70, 146]. The vertical vibrations are changing pressure in the disc-lining contact region. The maximal acceleration values of the support are reached in the vertical direction, while the response of the strut appears in the longitudinal direction. Such a situation occurs because the suspension form makes the support move vertically, initiating thereby rotation of the knuckle and simultaneously shifting the strut over the whole length.

Figure 6.24a shows acceleration records of the strut and support. The reaction of the strut of 39 Hz frequency agrees well with perturbing vibrations of the support.

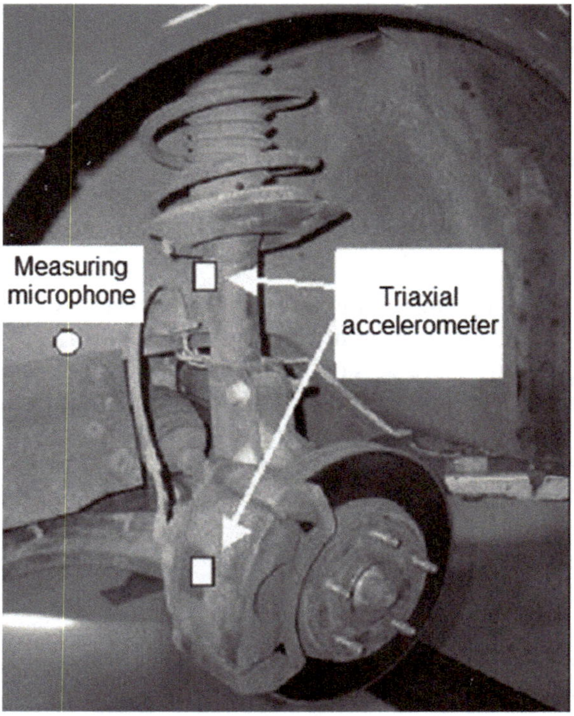


**Fig. 6.23** A general view of McPherson's suspension fit with a microphone and accelerometers [70]

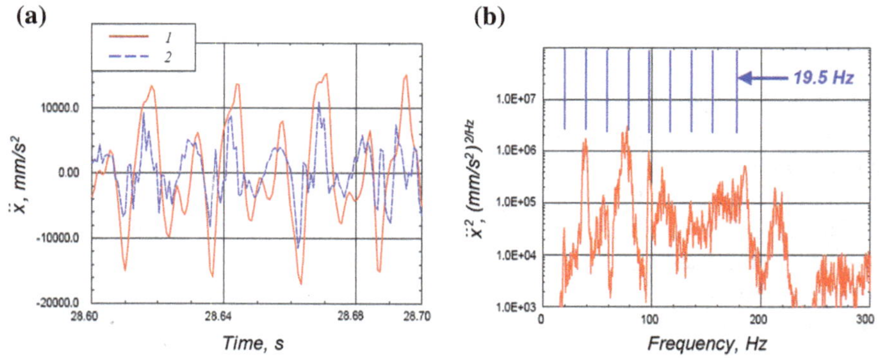


**Fig. 6.24** Measurement results of groan vibrations: time (**a**) and frequency (**b**) dependences (scale 0.1 s) of acceleration of the strut lengthwise (*1*) and that of the support endlong (*2*)

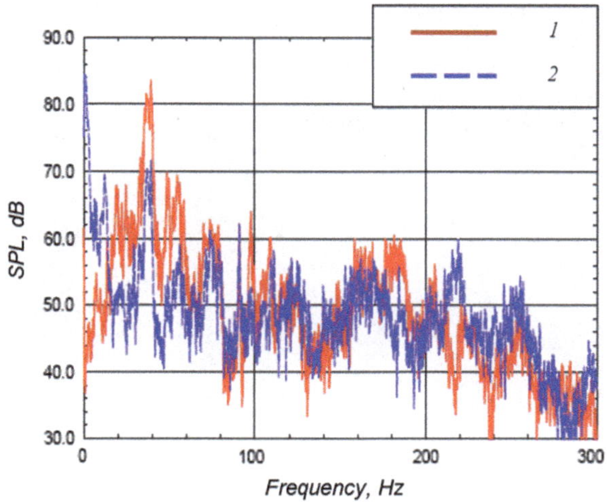


**Fig. 6.25** The internal (*1*) and external (*2*) noise levels during braking with groan

The spectral density diagram of the acceleration power of the strut presents spectra of the groan (Fig. 6.24b). The dominating spectral filling is observed at 39 and 77 Hz frequencies. These responses are observed together with their harmonics at higher frequencies multiple to the order of 39 Hz. There also occur but to a lesser degree the responses of a half order at a frequency 19.5 Hz. In principle, it is characteristic of the groan to have a great number of harmonics in its spectrum. The first harmonic (order) of the strut of McPherson's suspension arises usually within the range 20–50 Hz.

Spectral density of the recorded sound pressure level (SPL) is shown in Fig. 6.25. The peak of this characteristic is seen at 39 Hz, which is considered as a most clearly perceived sound of groan. Simultaneous measurements by a few microphones have unambiguously proved that the SPL is higher inside the car cabin as compared to the outside sound near the support. This confirms the structural origin of the groan effect. Besides, this type of the brake noise may be intensified by the interior acoustics.

It is supposed that the noise initially passes through the strut fixture to the cabin. This is proved by the agreement between a strong spectral filling of acceleration of the strut and the interior SPL at 39 Hz frequency.

The spectra of acceleration of the strut and the interior SPL have shown that the vibration transferability reduces significantly at above 60 Hz. This in part occurs due to the loss of the transmitting power in the strut fastening point or because of a stronger effect of transferability from the place of fastening till the driver's right ear. This fact is confirmed by Fig. 6.26 illustrating a typical partial amplitude–frequency characteristic for the place of strut fastening. The response shows that the transmitting power drops by 10 dB as the frequency rises from 40 to 80 Hz.

Evidently, the reduction of groan vibrations lowers the corresponding noise. By achieving transfer characteristics, we can adjust the groan vibration frequency so as to match it to the minimal transmitting power, thus hampering its spreading in the

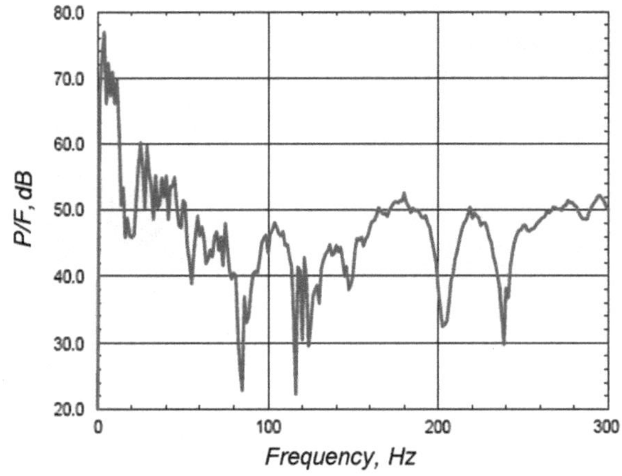


**Fig. 6.26** Amplitude-frequency transfer characteristic (force *F* in the point of strut fastening versus sound pressure *P* in the car cabin)

cabin. The minimal amplitude–frequency characteristic at about 90 Hz shown in Fig. 6.26 defines the groan frequency, providing thereby the corresponding noise reduction.

## 6.5.2 Theoretical Investigations of Brake Groan

Different models have been developed to simulate friction dynamics and study design effects on groan characteristics. These models are based either on experimental data concerning the static and dynamic coefficients of friction as well as theoretical results of the friction processes [69, 70, 147]. To characterize the friction force leading to a stick-slip motion, it will be helpful to analyze the nonstationary nonlinear dynamics. For instance, the brake groan has been analyzed in [70] by a MSC.ADAMS software package extensively used today to creation such models. A corresponding model is shown in Fig. 6.27 for a suspension and a brake system. Since the suspension with a McPherson strut is independent, it seemed rational to simulate a quarter of the car.

The main visible components of the suspension model are a steering arm, a journal, strut, disc, support assembly and a tire (wheel).

Although the outline of the car body is not seen in Fig. 6.27, a rigid concentrated mass presenting the body is included into the model.

It is anticipated that some components may display elastic properties in the typical for groan frequency range (200–500 Hz), so they cannot be presented as rigid ones. The MSC.ADAMS software package gives a possibility to present the elastic components as the elastic bodies, and their geometry can be imported from a package of the finite-element analysis, like MSC.NASTRAN or I-DEAS. Table 6.3 illustrates how each component is presented in the model of the system.

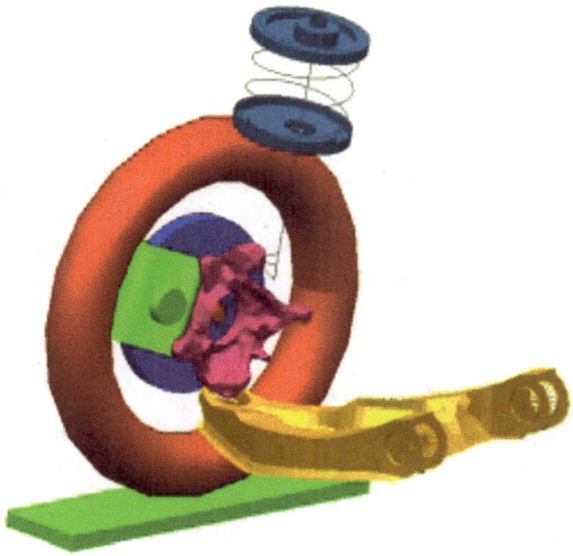

**Fig. 6.27** A suspension model with McPherson's strut for the brake groan studies [70]

**Table 6.3** Modeling of brake components

| Component | Model representation |
|---|---|
| Arm | Elastic body |
| Journal | Elastic body |
| Strut | Elastic body |
| Disc | Rigid concentrated mass |
| Support assembled | Rigid concentrated mass |
| Tire/wheel | Lumped parameter |
| Body | Rigid concentrated mass |

It seems most difficult to preset correctly the properties of the tire–wheel subsystem. The elastic properties of the tire and vibrations do exist in the considered frequency range. In the finite-element model the subsystem tire–wheel can be presented as an elastic body using the ADAMS software. It should be noted that such finite-element models are hard to construct in practice. In this connection, we have used instead the model of the lumped parameters (a set of springs and masses) to represent the tire–wheel subsystem. The characteristics of the springs and masses are chosen so as to impart stiffness to these components comparable to the results of the modal analysis of the system.

Simulation of the friction force is considered to be one of the most complex problems in theoretical studies of groan [73, 148]. The force acting between the disc and the pad can be described by a standard Amonton–Coulomb's equation of friction

$$F = \mu(\upsilon_r)N, \tag{6.1}$$

where $F$ is the friction force operating in tangential direction; $N$ is a normal force of brake pressure, and $\mu(\upsilon_r)$ is the friction coefficient dependent on relative velocity $\upsilon_r$ between the disc and the friction lining. To find the dependence of the friction coefficient on sliding velocity one may use different analytic expressions enabling to describe friction at a transfer from the condition of rest to sliding. The friction coefficient value in the vicinity of the null sliding velocity is usually described by the equation [149]:

$$\mu(\vartheta_r) = \frac{0{,}4}{\pi} arctg(200 \cdot \vartheta_r)\left(\frac{1}{|\vartheta_r| + 1} + 1\right), \tag{6.2}$$

A complicated dependence shown in Fig. 6.28 presents a special interest for groan simulation. In contrast to other models, this graph evidently dependents on its travel line according to the following reasoning. With increasing relative velocity (at transfer from sticking to sliding) the curve conforms to the standard statico-dynamic model of friction. Before transition to sliding, a higher coefficient of static friction is used, and while the transition to sliding occurs at slippage, a lower coefficient of dynamic friction is used. However, when the relative velocity drops at transition from sliding to slippage, this relation becomes improbable. The friction force increase at deceleration turns to be inadmissible either from the standpoint of physics. Therefore, for the case of the decreasing relative velocity an alternative path is taken, which enables to make the friction coefficient constant and equal to the dynamic one [70].

The computation results of the friction-excited self-vibrations in the automotive brakes fit with a McPherson strut are illustrated in Fig. 6.29 [70, 146–148]. The time dependences of sliding velocity and support acceleration in vertical direction are shown on the diagram in Fig. 6.29a. The initiation of sliding coincides with an

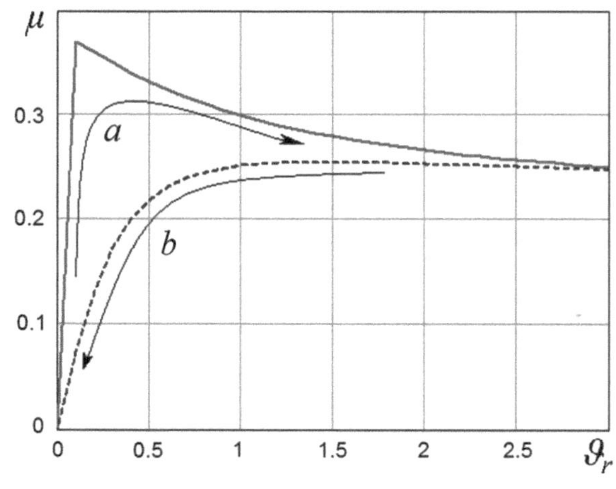


**Fig. 6.28** The friction coefficient versus relative velocity at transition from statics to sliding (**a**) and from sliding to rest (**b**)

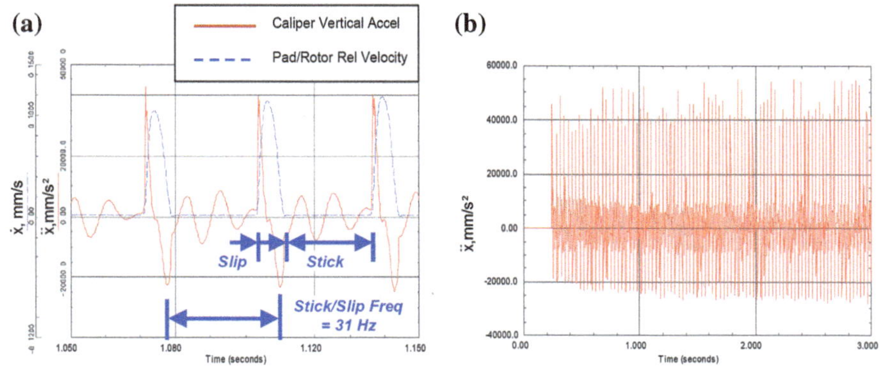


**Fig. 6.29** Simulation results of brake groan: *1* acceleration of support in vertical direction; *2* sliding velocity; time scale 0.1 s (**a**), 3.0 s (**b**)

abrupt increase (a leap) in both acceleration (curve 1) and sliding velocity (curve 2). This stage terminates by decreasing of the velocity till zero and transfer to a rest. Figure 6.29b presents a time dependence of support acceleration (3 s) that characterizes the vibration load on the friction joint.

The frequency of repetitions of the "stick-slip" cycles (31 Hz) found theoretically is close to the experimental one (39 Hz) derived in the ride tests.

The calculated spectra of accelerations obtained by the FFT method are illustrated in Fig. 6.30 for the vertical acceleration of the support (Fig. 6.30a), longitudinal accelerations of the strut (Fig. 6.30b) and for the force *F* operating in the strut—car body junction (Fig. 6.30c). The analysis of the spectra presented in Fig. 6.30 has proved that most of the values are interrelated with the friction-induced self-vibrations in the brake unit since their frequencies are multiple to the main groan frequency (1st order). Similarly to the experiment, the spectra of longitudinal accelerations of the strut and vertical ones of the support coincide in fact and are analogous by their form to the force *F* spectrum.

It is especially important to calculate this force since its measurement is complicated in practice. Nevertheless, this joint of the strut with the car body is considered as a chief link of vibration propagation inside the car. In this connection, it is possible to estimate the force affecting the junction between the strut and the body in terms of acceleration of the strut only.

The theoretical and experimental results of the automotive suspension equipped by a McPherson strut have shown that the dynamics and characteristics of groan are connected with longitudinal vibrations of the tire and strut. It has been proposed to adjust the suspension design by its structural modification so as to make the main frequency (1st order) aligned with the minimal transfer amplitude-frequency of the strut—body junction. It has been established theoretically elsewhere [70] that stiffening of the journal—strut system elevates the vibration frequency and decreases its level. The reduction of stiffness of the spring and addition of mass on its coils influences favorably on hampering transmission of vibrations onto the car body.

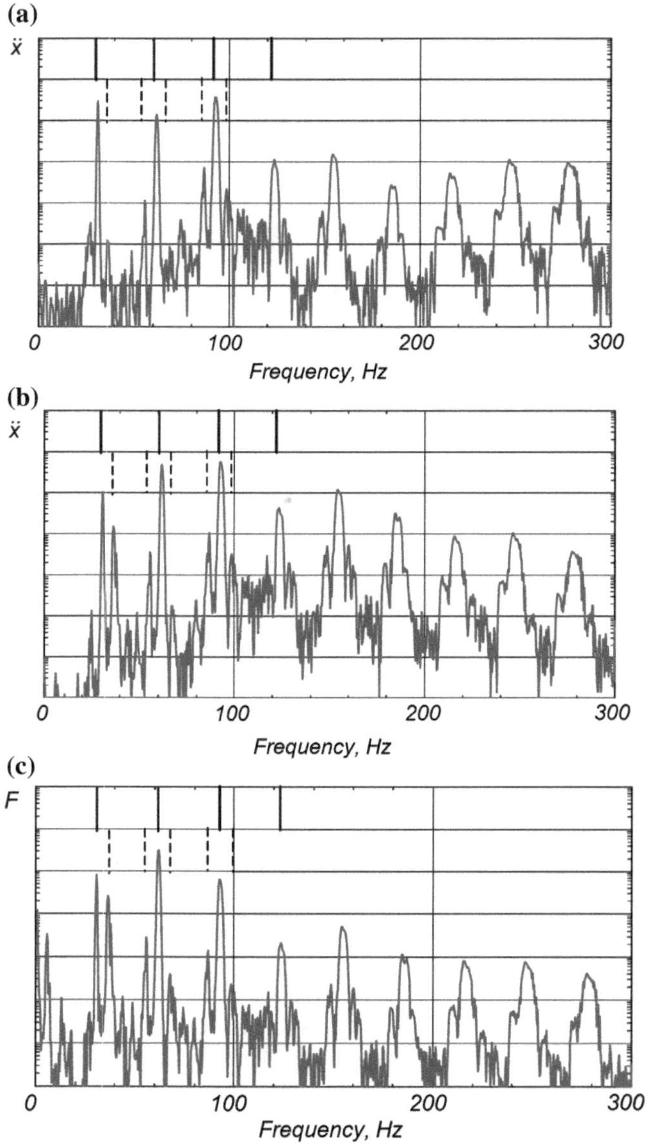


**Fig. 6.30** Simulation results of the brake groan: spectra of vertical acceleration of the support (**a**), of the strut (**b**), spectra of the forces in the strut-car body junction (**c**)

## 6.6 High-Frequency Acoustic Radiation in Brakes (Squeal)

The processes related to squeal and groan have been studied at length. The investigations in dynamics of nonlinear dissipative mechanical systems have generally confirmed for unstable states that the domain where several solutions exist decreases as damping of a system increases and the perturbation spectrum broadens [150, 151]. The investigation results published by different authors show that certain noise reduction can be reached by increasing Young modulus of the rubbing materials. Besides, there exists an important experimental evidence that squeal can be lowered via decreasing friction coefficient [49, 152–155]. However, inevitable reduction of the friction force affects adversely the efficiency of the brake system. Theoretical investigations dealing with the molecular dynamics of a nonlinear elastic finitely-extensible polymer model have proved that the propensity of a frictional material to generate squeal can be reduced by maintaining high and stable friction coefficient if to employ a polymer for the matrix phase with a certain set of structural properties [156].

Frictional interactions between the components of mechanical systems are commonly leading to high-frequency vibrations. Their nature can be related to both friction-induced nonlinearity and discreteness of the actual contact. Therefore, the vibrations may occur even in the idealized case when the static and dynamic coefficients of friction are found in a parity and their frequency is above 1 kHz in the supersonic range.

The experimental investigations [157] have established the following peculiarities under similar loading, sliding velocities and other frictional conditions:

- the constituents of vibroacoustic spectra conditioned by frictional processes are structurally dependent characteristics;
- the relative frequency and level of the friction-induced noise and vibration components are dependent on the magnitude and stability of the friction coefficient and damping capacity of the rubbing materials.

The HF vibration and noise research is extensively computerized. The mathematical models developed for different brake modifications are employed to study the dynamics of HF vibrations (squeal, groan) under nonstationary friction and to select structural parameters for the friction pairs and the whole joint [158–160]. The computing techniques are critical for the efficient try-off and optimization of different design variants in terms of their vibroacoustic parameters in order to avoid the expensive design errors.

### *6.6.1 The Methods of Analyzing Dynamics of Structures*

A traditional design method for dynamics of brake systems related to squeal is a modal analysis that includes estimation of the complex eigenvalues and spatial visualization of the vibration modes corresponding to all potentially unstable states of the system [161, 162]. Besides, the analytic methods of nonstationary processes are gaining popularity that in contrast to the modal analysis study transient processes with account of the time-dependent factors [15, 74, 163].

Since the braking systems and their elements are intricately shaped with many degrees of freedom, to realize each of above-named approaches it will be expedient to use the finite-element method [15, 49, 52, 59, 161–168].

*Modal analysis*. Along with above approaches, to compute the complex eigenvalues characterizing a vibratory system it is also possible to use the projective methods on the Krylov subspaces [169]. In a general case, motion of a system is described by the following inhomogeneous differential equation of the 2nd order [57, 59]

$$[M]\{\ddot{x}\} + [C]\{\dot{x}\} + [K]\{x\} = \{F_f\}, \tag{6.3}$$

where $[M]$, $[C]$ and $[K]$ are, correspondingly, matrix of the masses, damping matrix including mechanical losses on the internal friction in the materials, and a stiffness matrix of the system, $\{x\}$, $\{\dot{x}\}$, *and* $\{\ddot{x}\}$ are, correspondingly, vectors of displacements, velocities and accelerations; $\{F_f\}$—vector of friction forces operating between the disc and the pad. Without account of the friction forces the inhomogeneous (6.3) can be written in a homogeneous form [15, 170]

$$\left(\lambda^2[M] + \lambda[C]\right) + [K]\{\Phi\} = 0, \tag{6.4}$$

where $\lambda$ and $\{\Phi\}$—eigenvalues and corresponding eigenvectors that can be complex. To solve this equation relative to $\lambda$, the system is made symmetric by neglecting damping and asymmetric elements of the stiffness matrix $[K]$. The condition of symmetry makes possible to find $N$ number of eigenvectors $[\phi_1,\ldots,\phi_N]$. Then, the initial matrices are projected onto the subspace of $N$ eigenvectors

$$[M]^* = [\phi_1,\ldots\phi_N]^T[M][\phi_1,\ldots\phi_N], \tag{6.5}$$

$$[C]^* = [\phi_1,\ldots\phi_N]^T[C][\phi_1,\ldots\phi_N], \tag{6.6}$$

$$[K]^* = [\phi_1,\ldots\phi_N]^T[K][\phi_1,\ldots\phi_N]. \tag{6.7}$$

Equation (6.4) with account of (6.5)–(6.7) in a complex form will look like

$$\left(\lambda^2[M]^* + \lambda[C]^* + [K]^*\right)\{\Phi\}^* = 0. \tag{6.8}$$

Finally, the complex eigenvector of the initial matrix is

$$\{\Phi\} = [\phi_1,\ldots\phi_N]^T\{\Phi\}^*. \tag{6.9}$$

This algorithm is described in more detail in work [171].

The complex eigenvalues corresponding to an $n - \lambda_n$ can be presented as follows

$$\lambda_n = \alpha_n \pm i\omega_n, \tag{6.10}$$

where $\alpha_n$ is a real part of $\lambda_n$, indicating stability or instability of the system, $\omega_n$—imaginary part $\lambda_n$, expressing frequency of the $n$-th mode.

The generalized solution of the initial equation of motion of the brake system relative to displacements $x$ is of the form

$$x = Ae^{\lambda t} = e^{\alpha t}(A_1 \cos \omega t + A_2 \sin \omega t) \tag{6.11}$$

Based on this method we can find all complex eigenvalues related to unstable states of the system and requisite for squeal [56, 171]. The state of a system is defined by the sign of the real part of the complex eigenvalue. The unstable state corresponds to a positive real part $\lambda$. An additional parameter for determining state of a system is a decrement factor $h$, which negative value indicates instability of the system

$$h = -\frac{\alpha}{\pi|\omega|}. \tag{6.12}$$

The main purpose of the modal analysis is to find the decrement factor corresponding to the dominating unstable states in the system. Besides, the knowledge of unstable modes makes possible to choose one of the ways to abate instability, e.g. by displacing modal frequencies into the instability region via structural changes (stiffness, geometry) in the system [52, 172].

A refined method described in works [173, 174] is used to study different mechanisms of modal interactions leading to dynamic instability with account of the friction forces existing in the system. The right-hand side of (6.3) in a general form is

$$\{F_f\} = \mu(\{N_s\} + \{N_d\}), \tag{6.13}$$

where $\mu$ is the friction coefficient, $\{N_s\}$ and $\{N_d\}$ are, correspondingly, the static and dynamic normal forces. The problems on the complex eigenvalues discard the static normal force applied by the piston onto the friction pad. The dynamic normal force induced by vibration of the metal disc and friction pad is found as follows

$$N_d = K_s(x_{N,disc} - x_{N,pad}), \tag{6.14}$$

where $x_{N,disc}$ and $x_{N,pad}$ are displacements of the disc and pad in the normal direction, $K_s$—local contact stiffness. Thus (6.3), with account of the friction forces will take the form

$$[M]\{\ddot{x}\} + [C]\{\dot{x}\} + [K]\{x\} = \mu \cdot K_s[K_f]\{x\}, \tag{6.15}$$

where $[K_f]$ is a matrix of the effective stiffness dependent on friction between the lining and the disc. This matrix is asymmetric and relating the relative normal displacement with the tangential force.

The complex eigenvalue of the $n$-th vibration mode with account of the friction coefficient can be expressed by an approximate [57]

$$\lambda_n^2 = -\omega_n^2 - \mu \cdot K_s\left[\Lambda_f\right]_{nn} + \mu^2 \sum_{\substack{n=1 \\ k \neq n}}^{N} sign(\omega_n - \omega_k) f\left([\Lambda]_{nk}, [\Lambda]_{kn}\right) + O(3) \quad (6.16)$$

Equation (6.16) presents an approximate solution of (6.4). The calculation error of $\lambda_n^2$ is a third order infinitesimal one $O(3)$. The second and third members of (6.16) are conditioned by the presence of friction in the system and take the form

$$K_s\left[\Lambda_f\right]_{nn} \sim \left\{\Phi_{T,rel}\right\}_n^T \left\{\Phi_{N,rel}\right\}_n, \quad (6.17)$$

$$f\left([\Lambda]_{nk}, [\Lambda]_{kn}\right) \sim \left(\left\{\Phi_{T,rel}\right\}_n^T \left\{\Phi_{N,rel}\right\}_k\right)\left(\left\{\Phi_{T,rel}\right\}_k^T \left\{\Phi_{N,rel}\right\}_n\right), \quad (6.18)$$

where

$$\left\{\Phi_{T,rel}\right\}_n = \left\{\Phi_{T,disc} - \Phi_{T,pad}\right\}_n, \quad (6.19)$$

$$\left\{\Phi_{N,rel}\right\}_n = \left\{\Phi_{N,disc} - \Phi_{N,pad}\right\}_n. \quad (6.20)$$

Indices $T$ and $N$ are related to the tangential and normal directions, correspondingly. Instability of the system is defined by the third member of (6.16) that includes scalar products of the relative normal and tangential displacements between the two modes. The modes for which the scalar products of the eigenvectors in (6.18) are large, show a strong tendency towards merging or separating depending on the function sign. Proceeding from the fact that instability is a result of a partial approach between two modes, then the modes for which the third member is large in (6.16) are considered as potentially capable of interacting and generating instability. A refined method based on this criterion makes possible to elaborate recommendations necessary for modification of the brake system designs in order to eliminate or minimize modal interactions thus excluding the possibility of squeal generation [57, 175, 176].

The simulation procedures based on the modal design method display certain disadvantages. To simulate squeal, the braking process, which is nonstationary in reality, is substituted by a set of stationary stages, where the sliding velocity and contact pressure distribution are preset constant. The complex eigenvalue is derived at each stage. Its positive real part indicates the degree of instability and is related to the probability of squeal generation or noise intensity. These stages are automatically calculated by a modern software [170] of the finite-element simulation. Nevertheless, used in the given method linearization of nonlinearities enables to achieve a satisfactory calculation accuracy only in the assumption that the friction mode is stationary within at least a short time interval. Moreover, such a peculiarity of the nonstationary friction as the dependence of the friction material properties on the thermal effect duration, and some other are not taken into account. Besides, it turns improbable to estimate the generated noise level because the positive real part corresponding to the unstable mode indicates the rate of the vibration frequency increase but does not estimate its value [59].

Theoretically, the resonant amplitude may grow infinitely in the systems without damping (of the internal friction losses). In reality, there always exists the vibration energy dissipation in any system. An external source of perturbation possesses has a limited energy while the system elements display dissipative properties [177]. Therefore, the vibration amplitude of a given system increases till some limiting value, which can be derived by analyzing nonstationary processes or by the experimental modal analysis that is also required for verification of the design models [38, 59, 61].

Even if the modes have a sufficient frequency diversity, insignificant shifts in their frequency (less than 3 %) connected with inevitable deviations in a real brake system arising in manufacture or operation, always lead to a probable instability because of modal interactions [57, 175]. Sensitivity of a tribosystem to external conditions may also cause squeal generation at braking.

*The analysis of nonstationary processes*. To determine stability in the stationary or close to stationary friction problems, it is sufficient to perform a modal analysis. This method is, however, inapplicable for the nonstationary processes or the nonlinear systems.

Figure 6.31 illustrates an experimentally derived temporal vibration signal of a friction pair at a double braking with squeal.

The signal spectrum, i.e., its representation in the frequency region (Fig. 6.32) can be obtained by the fast Fourier transform applied for a time sampling related to squeal. In the case there are significant nonlinearities in the spectrum, the only reliable method for analyzing dynamic properties of this system will be the analysis of nonstationary processes [15, 74, 163, 178]. Theoretically, the analysis of nonstationary processes does not require any assumptions for the modal analysis model. These assumptions may include the constant contact area between the disc and the linings, the linear law of friction, and the dependence of the material properties on the time of braking [59]. The load variations affecting the time-dependent properties of

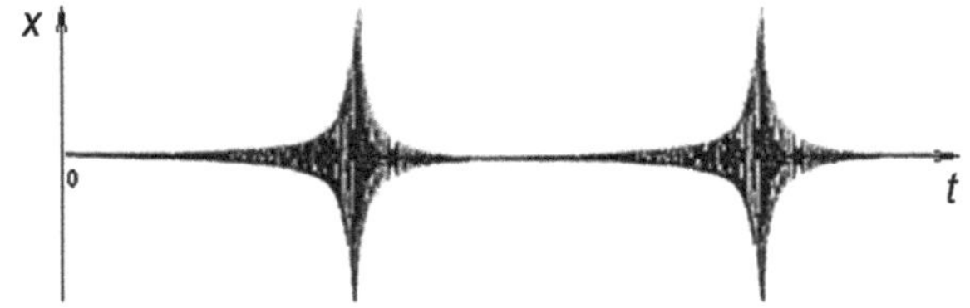


**Fig. 6.31** Vibrovelocity of a friction pair at a double braking with squeal [59]

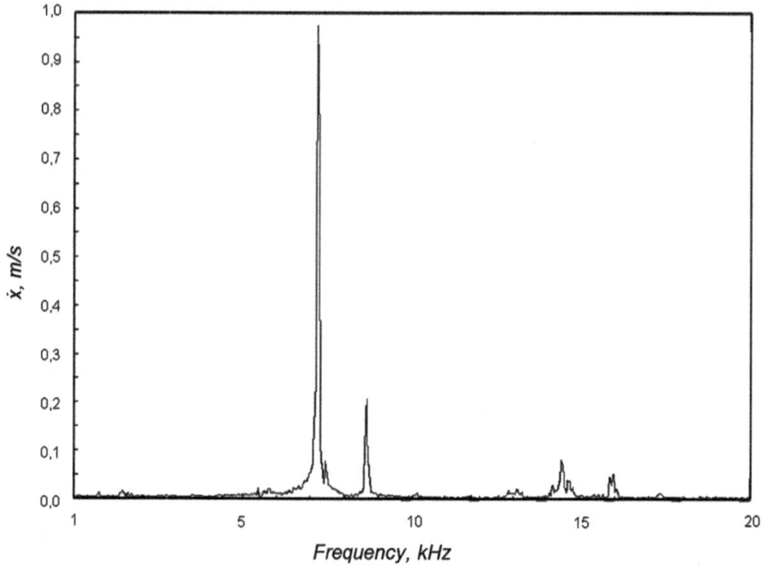


**Fig. 6.32** Vibrovelocity spectrum of a friction pair during brake squeal generation [179]

the material at braking can be accounted for in the nonstationary analysis with a high enough accuracy. The nonlinear nonstationary solutions are helpful in analyzing the effect of the nonstationary load and assessing stability. A nonstationary process can be analyzed at the initial stage till reaching an equilibrium (stationary) state. This makes possible to allow for the effect of other time-dependent factors. When a nonstationary process promotes the formation of a limiting cycle with the high-amplitude vibrations, the frequencies of these vibrations are commonly related to the brake squeal phenomenon [178].

Named method is realized in practice by the modern software packages of the finite-element simulation based on solution of the equation of motion and the rule of centered difference:

$$[M]\{\ddot{x}\}^{(t)} = \{F_{внеш}\}^{(t)} - \{F_{внутр}\}^{(t)} \tag{6.21}$$

where $[M]$—diagonal matrix of concentrated masses, $\{F_{внеш}\}$—vector of external forces, $\{F_{внутр}\}$—vector of internal forces, $t$—time. The relations for the vibro-velocity and vibrodisplacement are found from (6.19) as follows

$$\{\dot{x}\}^{(t+0,5\Delta t)} = \{\dot{x}\}^{(t+0,5\Delta t)} + \frac{\Delta t^{(t+\Delta t)} + \Delta t^{(t)}}{2}\{\ddot{x}\}^{(t)} \tag{6.22}$$

$$\{x\}^{(t+\Delta t)} = \{x\}^{(t)} + \Delta t^{(t+\Delta t)}\{\dot{x}\}^{(t+0,5\Delta t)} \tag{6.23}$$

where $\Delta t$—time increment per integration step.

Just as explicit [163, 180] so implicit methods [181] can be used at integration in the time domain. The finite-element method was for the first time used in work [180] t to analyze brake squeal with account of the nonlinear properties of the system and a nonstationary operation regime. It was established that stability of the system depends on the disc-pad contact properties. The contact dynamics was considered in terms of the *penalty* function and the principle of *virtual displacements* [163]. One of the advantages of work [163] is that the model employs the friction coefficient dependence on the contact pressure obtained experimentally in the course of the bench tests. To shorten the time of calculations in the explicit method of integrations keeping a desired accuracy, the authors have used a special type of the finite element with a reduced number of integration points [170]. The FFT of the computed signal has proved that some of the forecasted frequencies of a high-level noise agree well with the experimentally observed squeal. This method of the finite-element simulation of the HF noise in brakes has been further elaborated in works [74, 181–183], etc. The analysis of nonstationary processes was applied in works [184–186] to squeal investigations in aviation multidisc brakes.

The chief drawback of the analytic methods of nonstationary processes consists in protracted computations making the design procedure of the friction joints rather laborious. The nonstationary problem solution together with the design decision requires essential software resources. We can add to mentioned problems the elevated upper boundary of the frequencies being analyzed, to overcome which the sampling time interval in the explicit integration should be reduced [163, 180]. Usually, the choice of a time step $\Delta t$ in the explicit integration is based on the system linearity

$$\Delta t \leq l/c,$$

where $l$ is a characteristic length of the finite element; $c$—sonic speed.

In order to allow for the high frequency contribution the time step should be small enough. For instance, if the analysis demands to take into account the frequencies till 10 kHz, the $\Delta t$ should not exceed $10^{-5}$ s. In the case of the implicit integration, the levels of the high-frequency squeal components turn to be underestimated. The cause of this phenomenon has not been clarified as yet. We agree with the authors of [74], it is hard to define a single most adequate method to analyze squeal in brakes. It is worthwhile using the methods deciding a concrete target at different stages of problem solution and its different aspects. It will probably be justified in this respect to develop further and make above-described methods complementary [69].

### 6.6.2 Validity of Design Methods

A general disadvantage of design methods is rather bulky mathematical apparatus, which requires a substantial time input even in the presence of the modern computer-aided calculation procedures with attraction of the finite-element simulation means. When estimating validity of up-to-date computer models and the methods of noise and squeal effect simulation we presume that the key properties of the system in question have been determined experimentally. At a closer examination we understand the proximity and narrowness of the modern numerical methods. This is evident, first of all, from unsatisfactory agreement between the simulation and experimental data obtained in the real ride and bench tests.

*Representation of inertial and elastic properties.* The mathematical models describe the objects according to the labtest achievements in the mechanics of solids that involves the majority of physical parameters, geometry and properties of materials. However, there exists a tendency to neglect such phenomena as inhomogeneity and residual stresses connected with operation or inelastic properties affecting pretreatment. The derived modal density may be so high within the audible range that even slight differences in the models may bring about changes in the eigenfrequency order able to become a decisive argument in recognizing whether the squeal was actually generated or not [187].

*Damping properties of materials.* It is common knowledge that groan, squeal and moan occur as a result of frictional instability of the rubbing solids. To raise the design accuracy of solutions we should understand correctly damping properties of a system. Unfortunately, it is hard to measure experimentally or embody in the large-scale models the frictional interactions and damping properties of materials, especially their structural boundaries and junctions due to their diversity and stochastic nature of the processes. The simulation procedures of vibroacoustic phenomena on the frictional contact do not as a rule takes into account damping characteristics of the rubbing materials in a brake system but tend to present them in a simplified way (Fig. 6.33). In this case, damping properties of a frictional lining characterize only its vibration insulating properties from the positions of reducing vibration transferability from the source onto the structure. As a result, the tribological peculiarities of the contact dynamics, being the prime cause of instability,

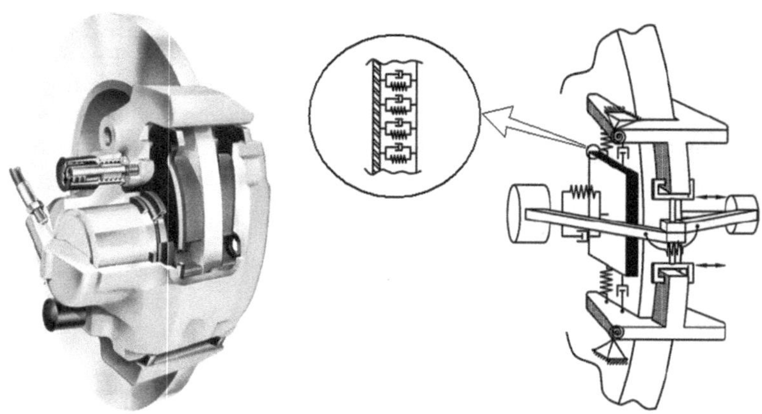

**Fig. 6.33** Dynamic scheme of automotive disc brake and representation of damping by a friction lining [190]

remain unaccounted. At the same time, it is acknowledged that even insignificant variations in the inlet parameters of damping may lead to considerable differences in the characteristics of stability [73]. This results in generation of a great number of unstable modes, although the experimental check of simulated objects has shown that squeal appears only on single frequency. To surmount this problem, it is possible to use a Rayleigh type of damping for at least some critical parts of the brake system [176, 188]. Representation of a large number of interfaces contributing to the structural damping may become a major problem by itself when using the methods of modeling and simulation [189].

*Gyroscopic effects*. Although squeal and moan appear under the low rotation velocities, at which gyroscopic forces are insignificant, it would be unjustified to neglect them because of the brake disc symmetry. The symmetry leads to formation of duplicating modes able to separate at a low rotation velocity. Recent investigations have proved that the gyroscopic effects should be taken into account for some types of flutter where the duplicating modes play a decisive role [191, 192].

*Friction models*. The numerical simulation of friction employs, most often the Amonton–Coulomb's model [70]. However, the choice of this model is not always justified because, first of all, the parameters of the model are unstable and nonstationary, which imparts indefiniteness to the system. A more promising model that takes into account the dependence of the sliding friction coefficient versus relative velocity is described in [193]. This kind of dependence is well known from the theory of lubrication as the Jercey-Stribeck diagram. Of interest is the fact that the negative gradient of the friction coefficient versus sliding velocity can theoretically lead to destabilization that was earlier considered as the main cause of the brake squeal. This phenomenon is treated today as an aftereffect in contrast to dynamic instability of the whole structure induced by the modal interactions or flutter. Unfortunately, there is lack of information concerning the perfected laws of friction intended for the noise and vibration simulation [194].

*Contact stresses*. The intricacy of the laws describing the normal and tangential stresses in the rubbing bodies adds difficulties to their numerical representation. Most known mathematically substantiated approaches for contact problems are the *method of penalties* and the *Lagrangian* one It is, however, acknowledged that just as a single asperity, so a random surface roughness may lead to a complicated nonlinear behavior of the whole system [73].In this connection, the fundamental investigations have been performed to examine the contact properties on the micro- and meso-levels [188, 194, 195]. The aim of these investigations is to determine the macroscopic properties using time or spatial mean values based on the micro- and meso-scopic variables.

*Thermal effects*. Heat-induced variations in properties and strains of materials are neglected, as a rule, in groan, squeal and moan simulation procedures. From the other hand, the effect of the brake fluid in the heat releasing environment should be subjected to a thorough consideration. It is common knowledge that compressibility of the brake fluid can alter essentially with temperature growth at prolonged braking, which affect the dynamic properties of the brake. Investigation results presented in [179] describe the temperature effect o a series of parameters, including elasticity modulus, friction coefficient and other. Variability of the properties as a result of temperature fluctuations can be reflected also by stochastic methods [196]. Thermal effects can be theoretically allowed for like in simulation of the heat-induced vibrations in the nonstationary analysis. However, in practice, this involves calculation problems even if to use the numerical methods.

The experimental and theoretical investigations of the brake groan [69, 70, 147] have proved that together with tangential vibrations of the brake support, the conjugated parts are also experiencing vibrations in the vertical direction. The vertical vibrations exert, in their turn, a significant effect on the contact pressure distribution between the disc and the pad. Therefore it is important to study in both kinds of investigations not the brake system alone but in conjunction with dynamics of the related components, such as suspension elements, car body, mounting hardware, as well as wheels and tires.

It is to be emphasized that there is no as yet a clear answer to the question, how sticking transmits into sliding in the systems with a large contact area. According to available data, groan has always been related to the stick-slip motion of the rubbing components. Nevertheless, even the novel miniaturized model systems show other results [197, 198]. The low-frequency measurements have proved that the whole area of the lining may be brought into a stick-slip contact like a rigid body only in some specific conditions [73].

Actually, nonlinearity and asymmetry of the friction-excited self-oscillations are promoting noise and vibration containing numerous high harmonics in their spectra. As for groan, it is subjectively perceived as a chaotic noise. Unfortunately, we have scarce information on whether this randomness stems from the dynamic system able to generate chaotic dynamics of the stick-slip motion [111], or from the boundary waves corresponding to the local breakdown or some other reasons. So it is evident that the present-day knowledge on interrelation between friction-induced self-oscillations and acoustic phenomena is insufficient for achieving reliable simulation of the brake groan.

Theoretical works on forecasting vibroacoustic behavior of brake systems are highly actual today, the more so, the experimental methods are rather expensive and have limited potentialities in optimizing friction joints. The numeric solutions of the problems on prediction and minimizing noise and vibration in brakes can be substantiated and adequately derived in provision of the next conditions:

a systematic approach to the problem in question;
application of adequate methods for a given problem (thermal vibration, groan, moan, squeal, etc.);
thorough interpretation of the test and simulation results.

The computer-aided models based on numerical methods display a number of undisputable advantages but their accuracy and reliability of prediction are bounded due to, first of all, assumptions and simplifications used in the models. In this connection, both methods and models used in this domain need further elaboration and perfection. One of the main problems in simulation in this sphere arises from the lack of research on tribological aspects of vibroacoustic phenomena on the friction contact.

The comprehensive theoretical and experimental investigations, as well as testing of the brakes with different friction lining materials have supported the fact that the friction material properties influence the level of self-excited and forced vibrations in the brakes, and just as LF so HF noise.

## References


1. A.V. Chichinadze, R.M. Matveevsky, E.D. Brown, *Materials in Tribological Engineering of Nonstationary Processes* (Nauka, Moscow, 1986), p. 248
2. M.P. Aleksandrov, *Brakes for Lifting-and-Shifting Machines* (Mashinostroenie, Moscow, 1976), p. 560
3. G.E. Blokhin, *Aviation Friction Clutches* (Oborongiz, Moscow, 1955), p. 208
4. S.G. Borisov, On development of the basics of raising durability of friction joints. Proc. NATI **237**, 3–9 (1975)
5. A.V. Chichinadze, A.G. Ginzburg, in *The Design of Temperature Regimes for a Sliding Bearing Rubbing Over a Fresh Track*, ed. by V.M. Sinaiskii, E.A. Marchenko. Heat Dynamics and Simulation of External Friction (Nauka, Moscow, 1975), pp. 5–11
6. A.V. Chichinadze, V.M. Goryunov, in *Temperature Regime of Sliding Bearings Under High Velocities*, ed. by A.V. Chichinadze. Heat Dynamics of Friction (Nauka, Moscow, 1970), pp. 70–77
7. V.L. Basinyuk, *Dynamics, Lubrication and Noise of Gearing* (MPRI NASB, Gomel, 2006), p. 216
8. L.G. Krasnevsky, The role of science-consuming components in mechanical engineering. Mod. Des. Methods Mach. **1**(2), 4 (2004)
9. A.V. Chichinadze, *Design and Research of External Friction at Braking* (Nauka, Moscow, 1967), p. 230
10. Yu.N Drozdov, V.G. Pavlov, V.N. Puchkov, *Friction and Wear in Experimental Conditions* (Mashinostroenie, Moscow, 1986), p. 224
11. A.V. Chichinadze, A.G. Ginzburg, V.M. Goryunov et al., Determination of serviceability criteria of materials under high sliding velocities. J. Frict. Wear **2**(3), 479–494 (1981)
12. V.P. Sergienko, V.V. Zhuk, V.A. Leshchev, A.V. Kupreev, Ecologically safe frictional materials for tractor brakes and transmissions. Tractors Agric. Mach. **12**, 12–16
13. S.M. Borisov, *Air-Actuated Friction Clutch* (Mashinostroenie, Moscow, 1971), p. 184
14. S.M. Borisov, *Friction Clutches and Brakes of the Road-Construction Machines* (Mashinostroenie, Moscow, 1973), p. 185
15. A.R. AbuBakar, H. Ouyang, Complex eigenvalue analysis and dynamic transient analysis in predicting disc brake squeal. Int. J. Veh. Noise Vib. **2**(2), 143–155 (2006)
16. A.A. Dmitrovich, G.S. Syroezhko, in *Proceedings of the International Scientific Conference on "Sintered Frictional Materials, Powder Metallurgy and Protective Coatings in Engineering and Instrument-Making"* (Minsk, 2003), pp. 22–29
17. M.I. Zlotnik, I.S. Kaviyarov, *Transmissions of Modern Industrial Tractors* (Mashinostroenie, Moscow, 1971), p. 247
18. V.G. Inozemtsov, V.M. Kazarinov, V.F. Yasentsev, *Automatic Brake* (Transport, Moscow, 1981), p. 464
19. S.S. Kokonin, E.I. Kramarenko, A.M. Matveenko, *Fundamentals of Design of Aviation Wheels and Brake Systems* (MAI, Moscow, 2007), p. 264

20. F.K. Germanchuk, *Durability and Efficiency of Brake Devices* (Mashinostroenie, Moscow, 1973), p. 176
21. N.K. Myshkin, S.S. Pesetskii, V.P. Sergienko, State of the art and promises in adoption of new polymeric materials in "Belarus" tractor design, in *Proceedings of the International Scientific Conference on "Promises in Development of Belorussian Tractor Industry"*, Minsk, 2006, pp. 133–146
22. P.L. Mariev, A.A. Kuleshov, A.N. Egorov, I.V. Zyryanov, *Open-Mine Motor Vehicles: State of the Art and Prospects* (Nauka, St. Petersburg, 2004)
23. V.A. Balakin, V.P. Sergioenko, M.M. Zabolotsky, A.V. Kupreev, Yu.V Lysenok, Thermal design of multidisc oil-cooled brakes. J. Frict. Wear **25**(6), 585–592 (2004)
24. L. Rabiner, *Theory and Application of Digital Signal Processing* (Prentice Hall, NJ, 1975), p. 762
25. A.V. Chichinadze (ed.), *Fundamentals of Tribology* (Nauka, Moscow, 2001), p. 664
26. A.S. Akhmatov, *Molecular Physics of Boundary Friction* (Fizmatgiz, Moscow, 1963), p. 472
27. G.M. Flidlider, Thermal stresses and stability of "Thin" friction discs in unsteady friction regime. J. Frict. Wear **2**(6), 77–85 (1981)
28. V.P. Sergienko, M.Yu. Tseluev, S.N. Bukharov, Prediction of Thermal Conditions for Multidisc Oil-Cooled Brake of a Mining Truck. *SAE Paper,* 2010-01-1713 (2010)
29. A.G. Ginzburg, in *Design and Research of the Axial Force with Friction in Splines of Multidisc Brakes and Clutches*, ed. by A.V. Chichinadze. Friction and Wear of Frictional Materials (Nauka, Moscow, 1977), pp. 13–19
30. I.N. Zverev, S.S. Kokonin, I.I. Khazanov, in *Friction Effect in Splined Joints of Multidisc Systems on Operation Uniformity of Friction Clutches and Brakes*, ed. by A.V. Chichinadze. Heat Dynamics of Friction (Nauka, Moscow, 1970), pp. 137–145
31. P.A. Vityaz, V.I. Zhornik, A.S. Kalinichenko, V.A. Kukarenko,V.Ya. Kezik, Improvement of tribological properties of heavy-loaded tribojoints based on the use of carbon nanocomponents, in *Proceedings of the International Scientific Conference on "Powder Metallurgy: Advances and Problems"*, Minsk, 2005, pp. 219–222
32. L.V. Ballon, *Electromagnetic Rail Brakes* (Transport, Moscow, 1979), p. 136
33. A.I. Kolesnikov, *Improvement of the Railroad—Rolling Stock Relationship* (Marshrut, Moscow, 2006), p. 365
34. V.A. Balakin, V.P. Sergienko, *Thermal Design of Brakes and Friction Joints* (MPRI NASB, Gomel, 2000), p. 220
35. S.M. Borisov, *Air-Actuated Friction Clutches* (Mashinostroenie, Moscow, 1971), p. 184
36. H. Abendroth, Worldwide brake—friction material testing standards, challenges, trends, in *Proceedings of the 7th International Symposium Yarofri, Friction products and Materials*, 9–11 September 2008, Yaroslavl, 2008, pp. 140–150
37. H. Abendroth, B. Wernitz, The integrated test concept: Dyno-vehicle, performance-noise, B. *SAE Paper*, 2000-01-2774 (2000)
38. V. Vadari, M. Albright, D. Edgar, An introduction to brake noise engineering. Sound and Vibration [Electronic resource] (2006), http://www.roushind.com. Accessed 15 Sept 2006
39. R. Mowka, Structured development process in stages of OE-projects involving with Western European car manufacturer, in *Proceedings of the 5th International Symposium of Friction Products and Materials Yarofri*, Yaroslavl, 2003, pp. 228–232
40. Yu.M Pleskachevskii, V.P. Sergienko, Friction materials with polymeric matrix: promises in research, state of the art and market. Sci. Innov. **5**, 47–53 (2005)
41. G.F. Alekseev, in *On Friction and Wear of Tribopairs Under Vibrational Loads*. Theoretical and Applied Problems of Friction, Wear and Lubrication of Machines (Nauka, Moscow, 1982), pp. 8–16
42. S.N. Bukharov, Reduction of vibroacoustic activity of metal-polymer tribojoints in nonstationary friction processes. Summary of Ph.D. Thesis, 5 Feb 2004, MPRI NASB, Gomel, 2010, p. 24

43. Vibration in Machinery. Ref. Book in 6 vols, ed. by V.N. Chelomei (Mashinostroenie, Moscow, 1979), vol. 2. Vibration of Mechanical Nonlinear Systems, ed. by I.I. Blekhman (1979), p. 351
44. A.H. Gelig, G.A. Leonov, V.A. Yakubovich, *Stability of Nonlinear Systems with More than One Equilibrium State* (Nauka, Moscow, 1978), p. 400
45. M.F. Dimentberg, *Nonlinear Stochastic Problems for Mechanical Vibrations* (Nauka, Moscow, 1980), p. 368
46. J. Wallaschek, K.H. Hach, U. Stolz, P. Mody, A survey of the present state of friction modelling in the analytical and numerical investigation of brake squeal, in *Proceedings of the ASME Vibration Conference*, Las Vegas, 1999, pp. 12–15
47. I.V. Kragelskii, V. Gitis, *Friction-Induced Self-Oscillations* (Nauka, Moscow, 1987), p. 184
48. V.P. Sergienko, S.N. Bukharov, A.V. Kupreev, Noise and vibration in brake systems of vehicles. Part 1: experimental procedures (review). J. Frict. Wea **29**(3), 234–241 (2008)
49. I. Ahmed, On the analysis of drum brake squeal using finite element methods technique. *SAE Paper,* 2006-01-3467 (2006)
50. N. Millner, An analysis of disc brake squeal. *SAE Paper*, 780332 (1978)
51. A. Bakar, H. Ouyang, J. Siegel, Brake pad surface topography Part I : contact pressure distributions. *SAE Paper*, 2005-01-3941 (2005)
52. G. Liles, Analysis of disc brake squeal using finite element methods. *SAE Paper*, 891150 (1989)
53. S. Kim, S. Cho, T. Yeo, A study on the effects of piston and finger offset on the pressure distribution at disk brake pad interface. *SAE Paper*, 2005-01-0794 (2005)
54. J. Fieldhouse, An investigation of the pad/disc dynamic centre of pressure using a 12 piston opposed caliper. *SAE paper*, 2007-01-3960 (2005)
55. J. Brecht, K. Schiffner, Influence of friction law on brake creep-groan. *SAE Paper* 2001-01-3138 (2001)
56. H. Ouyang, J.E. Mottershead, Friction-induced parametric resonances in disc: effect of a negative friction-velocity relationship. J. Sound Vib. **209**(2), 251–264 (1998)
57. M. Donley, C.H. Chung, Mode coupling phenomenon of brake squeal dynamics. *SAE Paper,* 2003-01-1624 (20030
58. S.-W. Kung et al., Modal participation analysis for identifying brake squeal mechanism, in *Proceedings of the 18th Brake Colloquium*, San Diego, CA, pp. 75–79, 1–4 Oct 2000
59. H. Ouyang, W. Nack, Y. Yuan, F. Chen, Numerical analysis of automotive disc brake squeal: a review. Int. J. Veh. Noise Vib. **1**(3/4), 207–231 (2005)
60. Brake Noise, Vibration, and Hardness: Technology Driving Customer Satisfaction [Electronic resource], http://www.akebonobrakes.com
61. W. Liu, J. Pfeifer, Introductions to brake noise & vibration. Honeywell Friction Materials [Electronic resource], http://www.sae.org/events/bce/honeywell-liu.pdt
62. W. Nack, A. Joshi, Friction induced vibration: brake moan. *SAE Paper*, 951095 (1995)
63. A. Gugino, J. Janevic, L. Fecske, Brake moan simulation using flexible methods in multibody dynamics. *SAE Paper*, 2000-01-2769 (2000)
64. M. Abdelhamid, Creep groan of disc brakes. *SAE Paper* 951282 (1995)
65. A. Crowther, J. Yoon, R. Singh An explanation for brake groan based on coupled brake-driveline system analysis. *SAE Paper*, 2007-01-2260 (2007)
66. H. Inoue, Analysis of brake judder caused by thermal deformation of brake disc rotors, in *Proceedings of the 21st FISITA Congress*, Belgrade, 1986, pp. 213–219, paper 865131
67. Kao T.K., Richmond J.W., Moore M.W. The application of predictive techniques to study thermo-elastic instability of brakes. *SAE Paper*, 942087 (1994)
68. H. Jacobson, Aspects of disc brake judder. Proc. Inst. Mech. Eng. Part D: J. Automob. Eng. **217**, 419–430 (2003)
69. J. Brecht, Mechanisms of brake creep groan. *SAE Paper*, 973026 (1997)
70. Donley M., Riesland D. Brake groan simulation for a McPherson strut type suspension. *SAE Paper*, 2003-01-1627 (2003)
71. M. Gouya, M. Nishiwaki, Study on disc brake groan. *SAE Paper*, 900007 (1990)

72. K. Popp, P. Stelter, Stick-Slip vibrations and chaos. Philos. Trans. R. Soc. **332**, 89–105 (1990)
73. N. Hoffmann, L. Gaul, Friction induced vibrations of brakes: research fields and activities. *SAE Paper*, 2008-01-2579 (2008)
74. Y.-K. Hu, A. Mahajan, K. Zhang, Brake squeal DOE using nonlinear transient analysis. *SAE Paper*, 1999-01-1737 (1999)
75. N.F. Kunin, G.D. Lomakin, Soundless dry external friction of metals at low velocities. J. Tech. Phys. **24**(8), 1361–1366 (1954)
76. N.F. Kunin, G.D. Lomakin, On interrelation between static and kinetic friction. Tech. Phys. **24**(8), 1367–1370 (1954)
77. K. Okayama, H. Fujikawa, T. Kubota, K. Kakihara, A study on rear disc brake groan noise immediately after stopping. *SAE Paper*, 2005-01-3917 (2005)
78. A. Crowther, R. Singh, Identification and quantification of stick-slip induced brake groan events using experimental and analytical investigations. Noise Control Eng. **56**(4), 235–255 (2008)
79. M. Abdelhamid, Statistical analysis of brake noise matrices. *SAE Paper*, 973019 (1997)
80. Resolving Customers' Brake Noise Issues. Case study. Brüel & Kjær Sound & Vibration Measurement A/S [Electronic resource], http://www.bksv.com/pdf/ba0798.pdf
81. K. Cunefare, Investigation of disc brake squeal via sound intensity and laser vibrometry. *SAE Paper*, 2001-01-1604 (2001)
82. S. Mahajan, Y.-K. Hu, K. Zhang, Vehicle disc brake squeal simulations and experiences. *SAE Paper*, 1999-01-1738 (1999)
83. J. Thompson, A. Marks, D. Rhode, Inertia simulation in brake dynamometer testing. *SAE Paper*, 2002-01-2601 (2002)
84. J.D. Fieldhouse, C. Beveridge, Comparison of disc and drum brake rotor mode movement, in *Proceedings of the Seoul 2000 FISITA World Automotive Congress*, Seoul, 2000, pp. 241–246
85. V.P. Sergienko, N.K. Myshkin, S.N. Bukharov, O.S. Yarosh, Investigations of the effect of friction material composition on vibroacoustic activity of tribojoints, in *Proceedings of the International Science Conference "Actual Problems of Tribology"*, 6–8 June 2007, Samara (Russia), 2007, pp. 266–278
86. J. Thompson, *Brake NVH: Testing And Measurements* (SAE International, London, 2011), p. 156
87. Automotive Brake NVH Dynamometer [Electronic resource], http://www.linkeng.com/. Accessed 25 Feb 2012
88. T. Hodges, Development of refined friction materials, in *Proceedings of the 5th International Symposium of Friction Products and Materials*, Yaroslavl, 2003, pp. 203–208
89. J. McDaniel, J. Moore, C. Shin-Emn, Acoustic radiation models of brake systems from stationary LDV measurements, in *Proceedings of the International Mechanical Engineering Congress and Exposition*, 1999, pp. 1–8
90. M. Fischer, K. Bendel, Hot on the trail of squealing brakes. *LM INFO*. Spec. Issue 1/2004, 9
91. J. Schell, M. Johansmann, M. Schüssler, D. Oliver, Three dimensional vibration testing in automotive applications utilizing a new non-contact scanning method. *SAE Paper*, 2006-01-1095 (1995)
92. R. Krupka, A. Ettemeyer, Brake vibration analysis with three-dimensional pulsed ESPI. Exp. Tech. **25**(2), 38–41 (2001)
93. R.R. Jones, C. Wykes, *Holographic and Speckle Interferometry*, Ch. 3–5 (Cambridge University Press, Cambridge, 1989)
94. J.D. Fieldhouse, P. Newcomb, The application of holographic interferometry to the study of disc brake noise. *SAE Paper*, 930805 (1993)
95. Z. Wang, A. Ettemeyer, Pulsed ESPI to solve dynamic problems. Application Report no. 04-97. Dr. ETTEMEYER GmbH & Co., 1997
96. G. Pedrini, B. Pfister, H.J. Tiziani, Double pulse electronic speckle interferometry. J. Mod. Opt. **40**, 89–96 (1993)

97. G. Pedrini, H.J. Tiziani, Double pulse electronic speckle interferometry for vibration analysis. Appl. Opt. **33**, 7857–7863 1994
98. J. Flint, J. Hald, Traveling waves in squealing disc brakes measured with acoustic holography. *SAE Paper*, 2003-01-3319 (2003)
99. Y. Desplanques, G. Degallaix, Interactions between third-body flows and localization phenomena during railway high-energy stop braking. *SAE Paper*, 2008-01-2583 (2008)
100. H. Jacobsson, Frequency sweep approach to brake judder, part A: the brake judder phenomenon. Classification and problem approach. Licentiate thesis, Chalmers University of Technology, Göteborg, 1998
101. T.K. Kao, J.W. Richmond, A. Douarre, Brake disc hot spotting and thermal judder: an experimental and finite element study. Int. J. Veh. Des. **23**(3/4), 276–296 (2000)
102. M.K. Abdelhamid, Brake judder analysis: case studies. *SAE Paper*, 972027 (1997)
103. W. Kreitlow, F. Schrödter, H. Matthäi, Vibration and 'hum' of disc brakes under load. SAE Trans. Sect. **1**, 431–437 (1985)
104. K. Augsburg, H. Brunner, J. Grochowicz, Untersuchungen zum Rubbelverhalten von Pkw-Schwimmsattelbremser. *Automobiltechnische Zeitschrift*, 1999, p. 101
105. W. Stringham, P. Jank, J. Pfeifer, A. Wang, Brake roughness—disc brake torque variation, rotor distortion and vehicle response. *SAE Paper*, 930803 (1993)
106. E.I. Marukovich, A.P. Markov, O.Yu. Bondarev, *Remote Flaw Detection on Contour Surfaces* (Belorusskaya Nauka, Minsk, 2011), p. 330
107. A.E. Anderson, R.A. Knapp, Hot spotting in automotive friction systems. Wear **135**, 319–337 (1990)
108. J.R. Barber, Thermoelastic instabilities in the sliding of conforming solids. Proc. R. Soc. Ser. A **312**, 381–394 (1969)
109. J.W. Richmond, T.K. Kao, M.W. Moore, in *The Development of Computational Analysis Techniques for Disc Brake Pad Design*, ed. by D.C. Barton, Advances in Automotive Braking Technology (MEP Ltd., London and Bury St. Edmunds, 1996), p. 158
110. R. Avilés, G. Hennequet, A. Hernández, L.I. Llorente, Low frequency vibrations in disc brakes at high car speed. Part I: experimental approach. Int. J. Veh. Des. **16**(6), 542–555 (1995)
111. T. Steffen, R. Bruns, Hotspotsbildung bei PkwBremsscheiben. *Automobiltechnische Zeitschrift*, 1998, pp. 100, 408–413
112. H. Jacobsson, Analysis of brake judder by use of amplitude functions. *SAE Paper*, 1999-01-1779 (1999)
113. H. Abendroth, T. Steffen, W. Falter, R. Heidt, Investigation of CV rotor cracking test procedures, in *Brakes 2000, International Conference on Automotive Braking-Technologies for the 21st Century*, London, 2000, pp. 149–162
114. D. Eggleston, Thermal Judder. *EURAC Technical Bulletin* 00034056 [Electronic resource]. http://www.eurac-group.com/documents/thermaljudder.doc
115. V.P. Sergienko, S.V. Bukharov, A.V. Kupreev, Tribological processes on contact surfaces in oil-cooled friction pairs. Proc. NAS Belarus **51**(4), 86–89 (2007)
116. D.D. Sterne, Monitoring brake disc distortions using lasers, in *Proceedings of the Institution of Mechanical Engineers, AUTOTECH'89 Conference*, 1989, IMechE paper C399/34
117. P.C. Brooks, D. Barton, D.A. Crolla, A.M. Lang, D.R. Schafer, A new approach to disc brake judder using a thermo-mechanical finite element model, in *Proceedings of the Institution of Mechanical Engineers, AUTOTECH'93 Conference*, 1993, IMechE paper C462/31/064
118. S. Gassman, H.G. Engel, Excitation and transfer mechanism of brake judder. *SAE Paper*, 931880 (1993)
119. D.G. Grieve, D.C. Barton, D.A. Crolla, J.K. Buckingham, Design of a lightweight automotive brake disc using finite element and Taguchi techniques. Proc. Inst. Mech. Eng. Part D: J. Automob. Eng. **212**, 245–254 (1998)
120. D. Thuresson, Thermomechanical analysis of friction brakes. *SAE Paper*, 2000-01-2775 (2000)

121. M.J. Haigh, H. Smales, M. Abe, Vehicle judder under dynamic braking caused by disc thickness variation, in *Proceedings of the Institution of Mechanical Engineers, Braking of Road Vehicles*, London, 1993, pp. 247–258, IMechE paper C444/022/93
122. M. Börjesson, P. Eriksson, C. Kuylenstierna, P.H. Nilsson, T. Hermansson, The role of friction films in automotive brakes subjected to low contact forces, in *Proceedings of the Institution of Mechanical Engineers, Braking of Road Vehicles*, London, 1993, pp. 259–267, IMechE paper C444/026
123. H. Jacobson, Wheel suspension related disc brake judder. *Proceedings of the ASME Design Engineering Technical Conference*, Sacramento, California, 1997, VIB-4165
124. A. De Vries, M. Wagner, The brake judder phenomenon. *SAE Paper*, 920554 (1992)
125. R. Bosworth, Investigations of secondary ride aspects of steering wheel vibration (shimmy and judder) using Taguchi methodology, in *Proceedings of the Institution of Mechanical Engineers, AUTOTECH 89*, London, 1989, IMechE paper C399/9
126. Engel H. G., Hassiotis V., Tiemann R. System approach to brake judder, in *Proceedings of the 25th FISITA Congress*, Beijing, 1994, vol. 1, pp. 332–339, paper 945041
127. M.K.Abdelhamid, Brake judder analysis using transfer functions. *SAE Paper*, 973018 (1997)
128. M. Kim, H. Jeong, W. Yoo, Sensitivity analysis of chassis system to improve shimmy and brake judder vibration on steering wheel. *SAE Paper*, 960734 (1996)
129. T.K. Kao, J.W. Richmond, M.W. Moore, Computational analysis of pad performance, in *Proceedings of the Institution of Mechanical Engineers, Braking of Road Vehicles*, 1993, paper C444/027/93
130. T. Valvano, K. Lee, An analytical method to predict thermal distortion of a brake rotor. *SAE Paper*, 2000-01-0445 (2000)
131. P.C. Brooks, D. Barton, D.A. Crolla, A.M. Lang, D.R. Schafer, A study of disc brake judder using a fully coupled thermo-mechanical finite element model, in *Proceedings of the 25th FISITA Congress*, Beijing, 1994, pp. 340–349
132. A. Floquet, M.-C. Dubourg, Realistic braking operation simulation of ventilated disc brakes. Trans. ASME J. Tribol. **118**, 466–472 (1996)
133. R.H. Martin, S. Bowron, Composite materials in transport friction applications, in *Brakes 2000, International Conference on Automotive Braking—Technologies for the 21st Century*, London, 2000, pp. 207–216
134. A.R. Daudi, W.E. Dickerson, M. Narain, Hayes' increased air flow rotor design, in *Proceedings of the Second International Seminar on Automotive Braking "Recent Developments and Future Trends"*, Leeds, UK, 1998, pp. 127–143
135. M.D. Hudson, R.L. Ruhl, Ventilated brake rotor air flow investigation. *SAE Paper*, 971033 (1997)
136. D.A. Crolla, A.M. Lang, Brake noise and vibration—the state of the art. Veh. Tribol. Tribol. Ser. **18**(Elsevier), 165–174 (1991)
137. R. Avilés, G. Hennequet, E. Amezua, J. Vallejo, Low frequency vibrations in disc brakes at high car speed. Part II: mathematical model and simulation. Int. J. Veh. Des. **16**(6), 556–569 (1995)
138. H. Jacobbson, Brake judder. Thesis, Chalmers University of Technology, Göteborg, 2001
139. R. Meyer, Brake Judder—analysis of the excitation and transmission mechanism within the coupled system brake, chassis and steering system. *SAE Paper*, 2005-01-3916 (2005)
140. A. Singh, G. Lukianov, Simulation process to investigate suspension sensitivity to brake Judde. *SAE Paper*, 2007-01-0590 (2007)
141. L. Zhang, D. Meng, Z. Yu, Theoretical modeling and FEM analysis of the thermo-mechanical dynamics of ventilated disc brakes. *SAE Paper*, 2010-01-0075 (2010)
142. F. Jardim, A. Tamagna, Study of the relationship between DTV, BTV and BPV over judder-type vibration of disc brake systems. *SAE Paper*, 2010-01-1694 (2010)
143. A.Yu. Ishlinskii, I.V. Kragelskii, On the friction-induced leaps. J. Tech. Phys. **14**(45), 276–282 (1944)
144. F. Bowden, L. Leben, The nature of sliding and the analysis of friction. Proc. R. Soc. Lond. Ser. A Math. Phys. Sci. **169**(938), 371–391 (1939)

145. E. Rabinowicz, The intrinsic variables affecting the stick-slip process. Proc. Phys. Soc. **71** (460 pt 4), 668 (1951)
146. V. Vadari, M. Jackson, An experimental investigation of disk brake creep-groan in vehicles and brake dynamometer correlation. *SAE Paper*, 1999-01-3408 (1999)
147. J.J. Xu, Disc brake low frequency creep groan simulation using ADAMS, in *Proceedings of the 2000 ADAMS International User Conference*, Orlando, Florida, June 2000
148. U. Kim, L. Mongeau, Simulation of friction-induced vibrations of window sealing systems. *SAE Paper*, 2007-01-2268 (2007)
149. G.P. Ostermeyer, On tangential friction induced vibrations in brake systems. *SAE Paper*, 2008-01-2850 (2008)
150. Yu.S. Pavlyuk, V.L. Sakulin, Nonlinear dynamics of mechanical systems under random effects [Electronic resource]. Chelyabinsk (2008). ftp://ftp.urc.ac.ru/pub/local/e-zines/DSM/v1_05.pdf
151. J. Wauer, J. Heilig, Dynamics and stability of a nonlinear brake model, in *Proceedings of the ASME* DETC 2001/VIB-21579, Pittsburgh, 2001
152. M.R. North, Disc brake squeal, in *Proceedings of the Conference on Brake of Road Vehicles, Institution of Mechanical Engineers*, 1976, C38/76, pp. 169–176 (1976)
153. T. Lewis, Analysis and control of brake noise. *SAE Paper*, 872240 (1987)
154. H.V. Chowdhary, A.K. Bajaj, C.M. Krousgrill, An analytical approach to model disc brake system for squeal prediction, in *Proceedings of the ASME DETC 2001/VIB-21560*, Pittsburgh, 2001
155. H. Ghesquiere, Brake squeal noise analysis and prediction. *ImechE Paper*, 925060 (1992)
156. M. Nishiwaki et al., A study on friction materials for brake squeal reduction by nanotechnology. *SAE Paper*, 2008-01-2581 (2008)
157. V.P. Sergienko, S.N. Bukharov, Vibroacoustic activity of tribopairs depending on the dynamic characteristics of their materials. Mech. Mach. Mech. Mater. **9**(4), 27–33 (2009)
158. F.R. Gekker, *Dynamics of machines with unlubricated operation of friction joints* (Machinostroenie, Moscow, 1983), p. 168
159. A.K. Pogosyan, V.K. Makaryan, G.S. Gagyan, Design of vibration stability of friction pairs in the disc-block brake systems of machines. J. Frict. Wear **12**(2), 225–231 (1991)
160. A.V. Chichinadze, A.D. Brown, F.P. Gekker, Modeling of friction joints on the example of multidisc aircraft wheels. Eng. J. **9**, 46–54 (1998)
161. S.W. Kung, K.B. Dunlap, R.S. Ballinger, Complex eigenvalue analysis for reducing low frequency squeal. *SAE Paper*, 2000-01-0444 (2000)
162. L.W. Chen, D. Ku, Stability of nonconservatively elastic systems using eigenvalue sensitivity. ASME J. Vib. Acoust. **116**(2), 168–172 (1994)
163. Y.K. Hu, L.I. Nagy, Brake squeal analysis by using nonlinear transient finite element method. *SAE Paper,* 971510 (1997)
164. H. Matsui, H. Murakami, H. Nakanishi et al., Analysis of disc brake squeal. *SAE Paper*, 920553 (1992)
165. M. Nishiwaki, H. Harada, H. Okamura et al., Study on disc brake squeal. *SAE Paper*, 890864 (1989)
166. W.V. Nack, Brake squeal analysis by finite elements. *SAE Paper*, 1999-01-1736 (1999)
167. G. Dihua, J. Dondying, A study on disc brake squeal using finite element methods. *SAE Paper*, 980597 (1998)
168. W.V. Nack, Brake squeal analysis by the finite element method. Int. J. Veh. Des. **23**(3–4), 263–275 (2000)
169. Y. Saad, Krylov subspace methods for solving large unsymmetric linear system. Math. Comput. **37**, 105–126 (1981)
170. ABAQUS Analysis User's Manual, Version 6.7. Dassault Systems (2007)
171. Q. Cao, H. Ouyang, M.I. Friswell, J.E. Mottershead, Linear eigenvalue analysis of the disc-brake squeal problem. Int. J. Numer. Methods Eng **61**, 1546–1563 (2004)

172. C.H. Chung, W. Steed, J. Dong, B.S. Kim, Virtual design of brake squeal. *SAE paper*, 2003-01-1625 (2003)
173. C.H. Chung, W. Steed, K. Kobayashi, H. Nakata, A new analysis method for brake squeal Part I: theory for modal domain formulation and stability analysis. *SAE paper*, 2001-01-1600 (2001)
174. H. Nakata, K. Kobayashi, M. Kajita, C. Chung, A new analysis approach for motorcycle brake squeal noise and its adaptation. *SAE paper*, 2001-01-1850 (2001)
175. N. Hoffmann, L. Gaul, Quenching mode-coupling friction induced instability using high frequency dither. J. Sound Vib. **279**, 471–480 (2005)
176. J.-J. Sinou, L. Jezequel, Mode coupling instability in friction-induced vibrations and its dependency on system parameters including damping. Eur. J. Mech. A **26**, 106–122 (2007)
177. A. Nashir, D. Jones, D. Henderson, *Damping of Oscillations* (Mir, Moscow, 1988), p. 448
178. H. Ouyang, W. Nack, Y. Yuan, F. Chen, On automotive disc brake squeal-part ii: simulation and analysis. *SAE Paper*, 2003-01-0684 (2003)
179. F. Chen, M.K. Abdelhamid, P. Blaschke, J. Swayze, On automotive disc brake squeal Part III: test and evaluation. *SAE Paper*, 2003-01-1622 (2003)
180. L.I. Nagy, J. Cheng, Y. Hu, A new method development to predict squeal occurrence. *SAE Paper*, 942258 (1994)
181. M.L. Chargin, L.W. Dunne, D.N. Herting, Nonlinear dynamics of brake squeal. Finite Elem. Anal. Des. **28**, 69–82 (1997)
182. Y. Chern, F. Chen, J. Swayze, Nonlinear brake squeal analysis. *SAE Paper*, 2002-01-3138 (2002)
183. H. Van der Auweraer, W. Hendricx, F. Garesci, A. Pezzutto, Experimental and numerical modelling of friction induced noise in disc brakes. *SAE Paper*, 2002-01-1192 (2002)
184. T.E. Rook, J.J. Enright, S. Kumar, R. Balagangsadhar, Simulation of aircraft brake vibration using flexible multibody and finite element methods to guide component testing. *SAE Paper*, 2001-01-3142 (2001)
185. O.N. Hamzeh, W.W. Tworzydlo, H.J. Chang, S.T. Fryska, Analysis of friction-induced instabilities in a simplified aircraft brake. *SAE Paper*, 1999-01-3404 (1999)
186. M.H. Travis, Nonlinear transient analysis of aircraft landing gear brake whirl and squeal, in *Proceedings of the ASME Design Engineering Technical Conference*, 1995, DE-vol. 84-1, vol. 3, Part A, pp. 1209–1216
187. A. Tuchinda, N.P. Hoffmann, D.J. Ewins, W. Keiper, Mode lock-in characteristics and instability study of the pin-on-disc system, in *Proceedings of the 19th International Modal Analysis Conference* (*IMAC-XIX*), Florida, 2001, pp. 71–77
188. K. Willner, Elasto-plastic normal contact of three-dimensional fractal surfaces using halfspace theory. J. Tribol. **126**, 28–33 (2004)
189. L. Gaul, R. Nitsche, The role of friction in mechanical joints. Appl. Mech. Rev. **54**, 93–106 (2001)
190. T. Jearsiripongkul, G. Chakraborty, P. Hagedorn, Stability analysis of a new model floating caliper disk brak. Integr. Comput. Aided Des. **11**, 77–84 (2004)
191. D. Hochlenert, G. Spelsberg-Korspeter, P. Hagedorn, Friction induced vibrations in moving continua and their application to brake squeal. J. Appl. Mech. **74**, 542–549 (2007)
192. O.N. Kirillov, Subcritical flutter in the acoustics of friction. Proc. R. Soc. A [Electronic resource]. http://onkirillov.narod.ru/2008-RSPA20080021.pdf
193. H. Hetzler, D. Schwarzer, W. Seemann, Steady-state stability and bifurcations of friction oscillators due to velocity-dependent friction characteristics. Proc. Inst. Mech. Eng. Part K J. Multibody Dyn **221**, 401–412 (2007)
194. G.P. Ostermeyer, M. Müller, H. Abendroth, B. Wernitz, Surface topography and wear dynamics of brake pads. SAE 2006-01-3202 (2006)
195. L. Gaul, R. Allgaier, W. Keiper, K. Willner, Untersuchungen zum Bremsenquietschen am Balken-Scheibe-Modell. VDI-Bericht **1736**, 17–31 (2002)

196. H.S. Qi, A.J. Day, K.H. Kuan, Forsala G.F, A contribution towards understanding brake interface temperatures, in *Proceedings of the* International Conference Braking, *2004*: *Vehicle Braking and Chassis Control*, 2004, pp. 251–260
197. S.M. Rubinstein, G. Cohen, J. Fineberg, Detachment fronts and the onset of friction. Nature **430**, 1005–1009 (2004)
198. S.M. Rubinstein, G. Cohen, J. Fineberg, Dynamics of precursors to frictional sliding. Phys. Rev. Lett. **22**(98), [Notes 226103.1–226103.4] (2007)

# Chapter 7
# Materials Science Approaches Towards Noise and Vibration Abatement in Nonstationary Friction Processes

The authors give technical characteristics and classifications of frictional materials for brakes and transmissions used in vehicles today. The investigations are reviewed intended to optimize structure and composition of friction materials for minimizing or eliminating self-oscillations, HF noise and LF judder. The principal parameters of frictional materials decisive in generating noise and vibration by the frictional materials in respective frequency ranges are analyzed. It is underlined that the approaches resting on the materials science standpoints are highly efficient in deciding noise and vibration reduction problems in tribosystems.

A great number of the methods are used in mechanical engineering to decrease or eliminate friction-induced self-vibrations and related vibroacoustic activity of frictional systems. They can be subdivided into two main groups: (i) the methods intended to improve tribological characteristics of tribopairs and (ii) the methods aimed at refining elastic and dissipative characteristics of the friction joints on the whole. Notice that the reduction of friction-excited self-vibrations till some level dependent mostly on structural features of the friction joint can be attained by both types of methods, while their total elimination is reached only by the first group of above-named methods [1, 2].

The LF acoustic vibrations in brake systems are usually damped by the sheet vibration-absorbing materials laid between the brake block and the brake cylinder face. The components of the HF noise are decreased by either changing characteristics of the thin-walled design elements or by application of the damping coatings on these elements, which reduces the resonant peak height [3]. It is possible to select an optimal coating in terms of calculated eigenfrequencies and modes or the experimental data on spectral characteristics of the vibratory and sonic radiation in the brake system [4]. We should, however, remember that any structural modification or coating application will inevitably raise cost of the ready items.

One of most promising directions in noise abatement in brake units is the development of a class of frictional materials possessing a stable friction coefficient, perfect thermophysical characteristics and a strong leap-preventing ability within the operating temperature and sliding velocity range [1, 2, 5–7].

The scientific data available for today evidence that above-named approach makes possible to select a required material for a tribopair at the design stage and



V.P. Sergienko and S.N. Bukharov, *Noise and Vibration in Friction Systems*,
Springer Series in Materials Science 212, DOI 10.1007/978-3-319-11334-0_7

thus influence its vibroacoustic characteristics. It is known that friction-excited self-vibrations contribute much to the noise and vibration effects in brakes. They are interrelated with the negative gradient of the friction coefficient and sliding velocity (meaning that the friction coefficient increases with decreasing sliding velocity) and its dependence on the stationary contact time. Nevertheless, the theories that explain their excitement by the processes occurring at a relative rest and motion of the rubbing bodies with account of respective staticokinetic characteristics are unable to define any acceptable practical criteria for the friction materials in order to estimate of a tribopair susceptibility towards vibroacoustic activity. Both static and kinetic characteristics of tribopairs depend on the load and velocity conditions, properties of the friction materials and many other factors. In a number of cases, friction-induced self-vibrations of the flutter type are feasible even in conditions of an ideally constant friction coefficient [8]. Moreover, it often turns impossible or hard to control staticokinetic friction characteristics of tribopairs when attempting to refine their vibroacoustic parameters because of the need to adhere to a preset complex of service characteristics, including frictional ones. The recent theoretical investigations [9] in molecular dynamics of a linearly elastic finitely tensile model of polymers have shown that the friction pair susceptibility to generate noise and vibration can be reduced in provision of maintaining high and stable friction coefficient by using a specific friction polymer with a certain set of structural characteristics for the matrix phase. There is, however, lack of scientifically grounded and systematized data on how the composition and structure of the rubbing bodies influence damping capacity, noise and vibration radiated by the tribojoints. This is in part connected with the difficulties in determining and processing a host of experimental data related to all ingredients of the material. Besides, there are scares investigation results on the real compositions of frictional materials since they constitute most often the manufacturer's know-how. The researchers have commonly to cite the results on the effect of separate ingredients in simple model mixtures [10–12].

Testing of the friction materials is a most expensive part of the design work, which shows a tendency to growth. The requirements in the juridical and warranty liabilities mainly connected with the noise and vibration problems at braking are toughening too [13, 14].

## 7.1 Classification and Technical Characteristics of Frictional Materials

Friction materials (FM) embrace a wide class of artificial materials designed to dissipate or transfer mechanical energy [15]. Friction materials are used in manufacture of brakes, friction discs of transmissions, clutch bushes, friction linings and dampers for mobile vehicles and rolling stock, in metallurgy, aircraft construction, engineering facilities in oil-production enterprises, railroads, transport systems, technological equipment and in many other machines and mechanisms. The areas

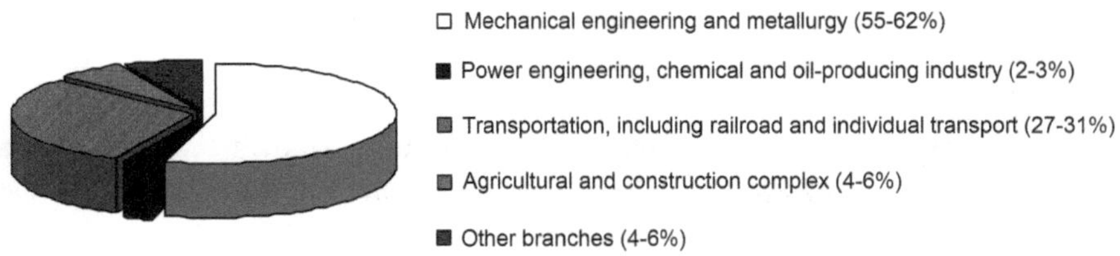


**Fig. 7.1** Consumption pattern of friction materials

of consumption of frictional materials on the polymer matrix base are shown in Fig. 7.1. Operation of the FM influences life quality of the population, first of all the transportation and production safety in which people are involved.

From the viewpoint of structural organization, friction materials with a polymer matrix present a composite matter consisting of different multiphase systems in which some organic polymer or a blend of polymers behaves like one of continuous phases (matrix) (Fig. 7.2). Their quality depends much on the wear resistance and thermomechanical strength of the matrix polymer. Friction composites contain reinforcing fillers in the form of commonly high-strength high-modular fibers that strengthen the polymeric matrix. There are also the components that improve thermophysical properties of the materials, thermal capacity and thermal conductivity first of all. They also include microsize dispersed friction modifiers in order to provide the desired tribological characteristics, structural plasticizers enabling to

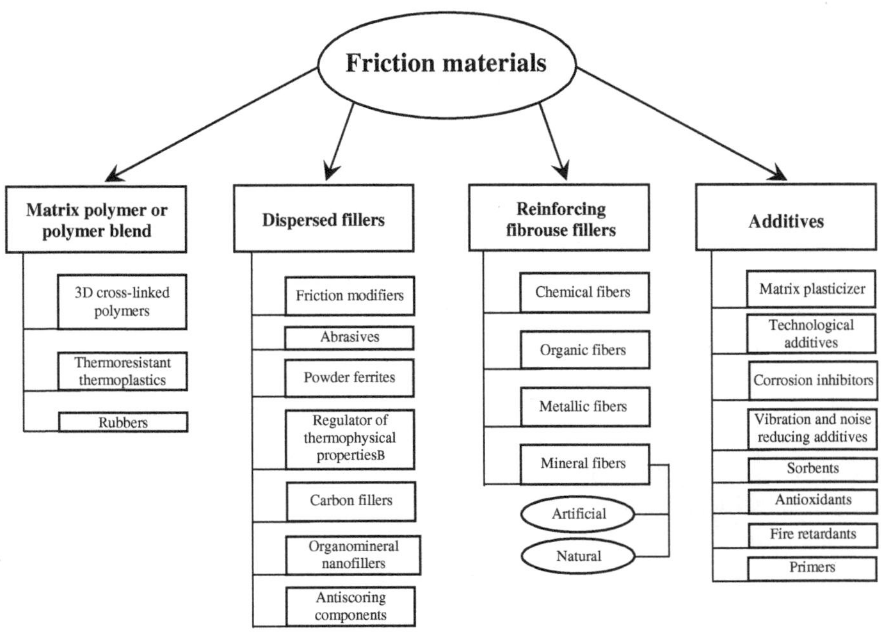


**Fig. 7.2** Structural organization of friction materials with a polymer matrix

reduce stiffness of the polymeric matrix, corrosion inhibitors, antiscoring additives and some other target dopes.

The FM may include about a hundred mineral, organic, synthetic or metal-containing substances in their commercial manufacture. Modern FM constitute a complex heterogeneous system composed of 12–40 ingredients that impart unique properties to the materials to make them perform in extreme service conditions [16].

Within the period of more than 80 years, during which the frictional materials science has gained strength, the main filler for the frictional polymeric compositions and some types of cermet materials was asbestos. From the viewpoint of frictional materials science, this natural material displays such a combination of unique properties that none of known in the art natural or artificial materials has. The term asbestos unites six different by their composition, structural-morphological and physico-chemical properties silicate fibrous compounds namely: crockidolite, anthophylite, actinolite, amosyte, tremolite, and chrysotile. The mineral chrysotile related to a group of magnesian hydrosilicates (theoretical formula $3MgO{\cdot}2SiO_2{\cdot}2H_2O$) is most often used for manufacture of the FM.

It has been established recently by a series of investigations that the fibrous 5–8 μm long and below 3 μm in diameter particles of asbestos present a biologically active substance with expressed carcinogenic affect. It is hazardous also because of a latent period between the onset of the negative affect of asbestos fibers and clinical signs of the illness lasting for till 30 years [17].

In 1982 a "Consolidated list of goods, which consumption and/or sale is prohibited, withdrawn, strictly limited or not adopted by the governments" was prepared proceeding from the decision of the general Assembly of the UNO. A complete group of asbestos materials is related to the extra hazardous for the man's health substances. Between 1982–1990s almost all means of using asbestos were prohibited in Western Europe, USA and many other countries. Application of asbestos-based materials is especially dangerous in the friction joints since its finest particles are accumulated in large amounts in the air round urban highways and in closed industrial areas. In spite of such strict measures adopted by the world community, there are in fact no limitations on the governmental level in CIS countries for the use of asbestos FM.

The domain of technical applications of the FM is restricted according to a series of requirements, such as provision of the friction coefficient stability and magnitude, independence of the sliding friction behavior of both service conditions and weather. The FM should perfectly run in to the counterbody without its wear and seizure, display enough mechanical strength, high wear resistance, desired thermophysical properties and high frictional heat resistance, corrosion stability, incombustibility, possess optimal vibroacoustic characteristics, ability to damp self-oscillations, ensure comfort at braking, and etc. In addition, the FM should be ecologically friendly, manufacturable, be cheap and supported by corresponding raw materials base. The FM are subjected to various adverse factors during operation, including high and sign-variable temperature and dynamic loads, high sliding velocities, intensive wear, often in the presence of hostile media (saline solutions, oils, acids, and other) [18].

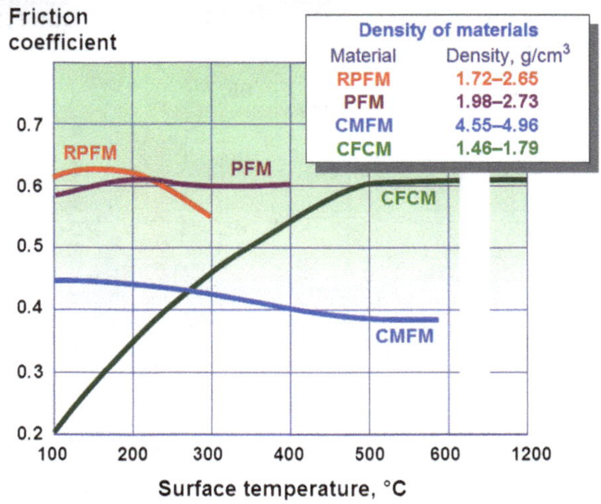


**Fig. 7.3** Comparative tribological characteristics of FM: *RPFM*—rubber-polymer FM, *PFM*—polymer FM, *CMFM*—cermet FM, *CFCM*—carbon frictional composite materials

By the present moment, three classes of the FM have remained in transportation and mechanical engineering: the composites based on organic matrix, metal-ceramic and carbon ones. Each of them offers a wide range of FM formulas that correspond to certain requirements and performances. The main criterion for selecting this or that material is heat parameters under which the tribopair is working. Figure 7.3 presents temperature intervals of the FM operation. The polymeric FM occupy the temperature range most relevant for modern engineering, wherefore their production volume reaches 90 % of the total production of frictional materials (Fig. 7.4).

The FM on the organic matrix base (RPFM and PFM) are limited in application within the volume temperature range $T_v$ = 300…400 °C and the mean friction surface temperature $T_s$ = 400…420 °C. Cermet FM are used under the heat conditions $T_v$ = 600…700 °C and $T_s$ = 800…1,000 °C, but their frictional efficiency drops as the temperature rises. Carbon-based FCM were developed for the operation in severe heat conditions like $T_v$ = 1,000…1,500 °C and $T_s$ = 1,500…2,000 °C and are used in homonymous tribopairs. Carbon FCM are inefficient at elevated environment humidity and low temperatures ($T_s$ < 330 °C), besides, they are rather expensive (400–1,200 $/kg.) A stable tendency is preserved for the last decades towards the production volume growth and consumption of the organic matrix-based FM (Fig. 7.4). Their efficiency is ensured by the exclusive properties, like high strength, frictional and vibroacoustic characteristics, manufacturability, durability and safety.

Vast experience has been accumulated in transport and engineering spheres concerning application of cermet friction items working in nonstationary friction conditions. Their advantages are manifested in the high specific friction power, heat cconductivity and wear resistance at elevated specific friction work. The specific friction power of cermet iron-based FM operating in the oil medium makes up 0.9–4.0 MWt/m$^2$, and admissible sliding velocity is till 80 m/s. The new

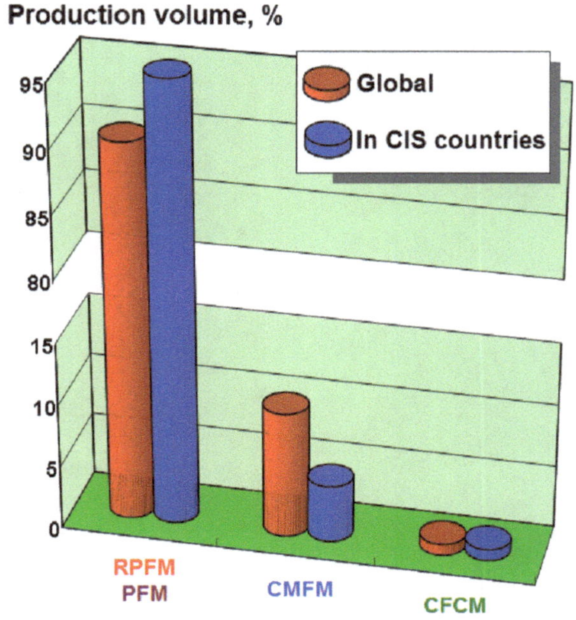


**Fig. 7.4** Production structure of friction materials

copper-graphite-based FM provide for the friction interaction efficiency at braking with a specific friction power till 6 MWT/m$^2$, and the specific friction power reaching 8.5 MJ/m$^2$ [19]. For comparison, the specific friction power of the paper FM is between 8.5 and 1.45 MWTm$^2$, that of the graphite-based FM is 0.5–1.5 MWTm$^2$, and the maximal efficient sliding velocity is 30–42 m/s. It is to be noted that cermet FM are inadequately operating in oil, i.e., they cannot ensure stability of the friction torque and smooth actuation of the friction joints, which impairs the dynamics of transient processes in the machines. Another important drawback is connected with a high vibroacoustic activity of tribopairs made of these materials. Unfortunately, it is hard to abate this tribological phenomenon in transmissions and brake systems, which impairs a subjective perception of the car quality [20, 21]. The present chapter cites the investigation results obtained when working on the problem of creating friction materials with improved vibroacoustic properties are discussed in.

## 7.2 Frictional Materials with Improved Vibroacoustic Characteristics

It is important to refine vibroacoustic characteristics of FM by observing safety and serviceability of the integral brake system. Along with the structural methods, there exist and are developing the ones based on materials science principles that are intended to inhibit vibroacoustic radiation in friction joints. For instance, the FM have been obtained by addition of the tung or flaxseed oil (0.5–10 vol.%) into the composition containing steel fibers, graphite, $BaSO_4$, phenol formaldehyde resins and powdery iron. Such a composition stabilizes the friction coefficient, enabling thereby to lower the level of noise and vibration [22].

To reduce noise at braking and increase stability of the friction coefficient, a polymeric molding composition is doped with butadiene styrene rubber containing 27–35 mass% of the nitriloacrylic acid chains, chopped brass wire, glass and carbon fibers [23].

Along with above-mentioned, the friction materials for operation under high temperatures have been developed. To these belong the FM having a decreased level of noise and wear, and improved stability of the friction coefficient [24]. Besides, the polymeric composites for friction purposes are available able to eliminate vibration of the brakes in conditions of high pressures and temperatures [25].

The rubbing bodies are anticipated to generate less pronounced groan if to manufacture the friction lining material by the following procedure. A special blend is prepared containing a graphite powder with some metal or alloy that is softer than steel. It is then ground into a powder and introduced into the binder with addition of either a steel powder or asbestos, glass or aluminum fibers. This blend contains: fibers 5–35 vol.%, a binder 10–35 vol.%, a metal 0.5–15 vol.%, and the rest is either graphite or organic fillers [26].

The brake squeal can be reduced by addition of such ingredients into the composition as kaolin, liquid glass, barium stearate, trisulfide antimony, fluorspar [27].

We should also mention a family of the porous frictional materials. The materials for frictional purposes with imparted high porosity are able to hinder noise generation and assist in dampening sonic waves, reducing thereby instability of the friction coefficient under elevated temperatures. One of such materials contains hollow carbon microspheres (15 vol.%), steel fibers (25 vol.%), metal oxides (10 vol.%), rubber (10 vol.%), graphite (10 vol.%), barium oxide (10 vol.%), and a thermosetting resin (20 vol.%). As a result, porosity of the material reaches 10–15 % [28]. For these aims microporous zeolites are often used as additions to increase porosity in frictional materials. It is to be noted, however, that the developed porous materials are decreasing the noise level in the friction joints, at the expense of reduced power parameters [29].

The vibration-absorbing coatings based on melamine varnish containing melamine formaldehyde resin and a modified alkyd resin dissolved in a mixture of aromatic hydrocarbons are extensively used today in various spheres [30].

Along with the coatings, we should mention the vibration–damping layers incorporated in composite materials that are widely used in industry. They display vibration-absorbing properties that ensure a logarithmic sound decay decrement 0.04–0.5. They contain 40–90 mass% iron oxide and 60–10 mass% binder (polyester resins, polypropylene, polyurethane, phenol formaldehyde resin, epoxide, acrylobutadiene nitrile rubber) intended for the noise level damping [31].

The following major approaches have been formulated proceeding from the analysis of scientific literature on friction materials with perfected vibroacoustic characteristics and corresponding investigations in the effect of structure and composition of frictional composites on their susceptibility to generate noise and vibration in tribopairs:

- Optimization of composition in terms of staticokinetic characteristics of friction. The essence of this method consists in maintaining growth of the kinetic friction coefficient dependence on sliding velocity and the minimal possible increase of the static friction coefficient in the stationary contact. Although in practice the difference between the static and kinetic friction coefficients is commonly minimized. This approach is recommended for hindering low-frequency friction-excited self-oscillations and noise (groan).
- Perfection of damping capacity. This method is based on increasing the dynamic elasticity modulus and the factor of mechanical losses (tangent of the angle of mechanical losses) in a given temperature interval. This approach is most efficient in reducing high-frequency (above 1 kHz) acoustic radiation (squeal).

The approaches aimed at reducing forced vibration. This direction is evidently most complex since a set of FM properties are optimized simultaneously to diminish uneven wear of the disc and its thermal buckling. Most important properties defining susceptibility of a tribopair to generate of low-frequency forced vibration is first of all its compressibility, "aggressiveness" (capacity to wear) relative to the counterbody, and thermophysical properties.

In the chapter to follow the use of above approaches is described in optimization of vibroacoustic characteristics in the form of examples of concrete investigation results of asbestos-free frictional composite materials designed for decreasing friction-excited self-vibrations (groan) and high-frequency noise in friction joints.

## 7.3 Optimization of Frictional Material Composition by Staticokinetic Characteristics of Friction

The effect of material formula on friction characteristics of a polymer composite incorporating 12 ingredients and interrelated groan generation in brakes is discussed in investigation results of works [10, 32, 33]. The formulas of the model composite materials (noncommercial) are presented in Table 7.1.

The friction composites were produced by mixing the initial components, preforming, thermocompression molding followed by thermal and mechanical treatment. In order to preserve shape of the brittle components and avert their thermal destruction, they were mixed in two stages. The preforming was carried out under 34.3 MPa pressure and 20 °C temperature. The hot pressing pressure was 31.6 MPa and 160 °C temperature during 10 min. The post-curing (thermal treatment) was conducted in a convection oven at 210 °C during 6 h. More detailed information on the manufacturing technology of commercial brake blocks can be found elsewhere [34]. The friction composites were subjected to tribological testing on a bench simulating a motorcar brake system operation. The friction parameters (contact

**Table 7.1** Formulas of model composite materials with a polymer matrix [10]

| Ingredients | | Content, vol.% | |
|---|---|---|---|
| | | Initial composition | Modified composition[a] |
| Matrix and other organic components | Phenol formaldehyde resin (PFR) | 10.0 | 8.0 |
| | Cashew powder | 10.0 | 12.0 |
| | Chopped rubber | 8.0 | 8.0 |
| Reinforcing fiber | Aramide fiber | 8.0 | 3.7 |
| | Steel fiber | 4.0 | 3.7 |
| | Mineral fiber | 10.0 | 14.6 |
| Abrasive particles and greases | $ZrSiO_4$ (zirconium) | 3.0 | 1.5 |
| | $Sb_2S_3$ | 3.0 | 3.5 |
| | Graphite | 10.0 | 11.0 |
| Other fillers | $BaSO_4$ | 25.0 | 25.0 |
| | $CaCO_3$ | 8.0 | 8.0 |
| | $Ca(OH)_2$ | 1.0 | 1.0 |
| | Total concentration | 100.0 | 100.0 |

[a] Modified composition is based on test results using a limited optimizing schedule

pressure, sliding velocity) and measurement results (friction force value, surface temperature of the brake disc) were recorded by a computer-aided system of data storage. To measure the temperature an IR pyrometer (3 M.Scotchtrak IR-16) was used. The working surfaces of each sample were preliminary ground till obtaining a uniform contact with the disc surface. The rubbing surfaces were run-in in the course of 50 brake cycles, each lasting 10 min. The kinetic friction coefficient was measured at a constant sliding velocity 6.92 m/s, contact pressure 0.687 MPa and initial surface temperature of the disc 100 °C. The static friction coefficient was measured by a high-precision torsion sensor under a similar contact pressure. The mean kinetic and static friction coefficient values were obtained on the base of 100 measurements. The main parameters used in the tribological test procedure [10] are listed in Table 7.2.

According to above table, the ingredients used in the friction composites under study can be subdivided into 4 groups depending on their purpose. The effect of the first three groups on the friction characteristics and interrelated friction-induced self-vibrations has been analyzed in work [10]. The effect of the fourth group is not considered since these ingredients influence but negligibly the friction characteristics. Thus, only nine components of three groups were examined. With this aim, 29 samples of the frictional materials were prepared of different compositions based on a *constrained mixture design* [35–37]. The volume content of each ingredient in

**Table 7.2** Parameters of tribological tests

| Stage | Contact pressure, MPa | Sliding velocity, m/s | Initial surface temperature, °C | Time of braking, min | Number of braking events |
|---|---|---|---|---|---|
| Run-in | 0.491 | 6.92 | 100 | 10 | 50 |
| Measurement of kinetic friction coefficient | 0.687 | 6.92 | 20 | 0.5 | 10 |
| Measurement of static friction coefficient | 0.687 | – | 20 | – | 10 |

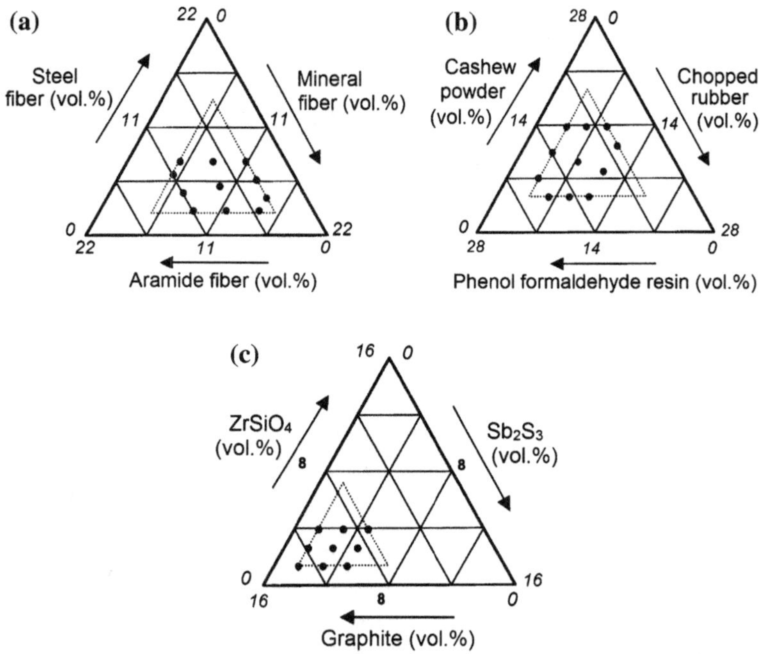


**Fig. 7.5** Experimental points for the constrained mixture design: reinforcing fibers (**a**), matrix and other organic components (**b**), abrasive particles and solid greases (**c**)

the composite was varied within the range ±50 %. A *simplex centroid design* was used to choose the experimental points. The results were analyzed in a canonical form by constructing polynomials that describe the results obtained on a corresponding face. The polynomial of the second order has been used in [10] for three variables. The experimental points for the constrained mixture design are shown in Fig. 7.5. The apexes of the simplex centroid design of the third order correspond to the total concentration of the components in each group.

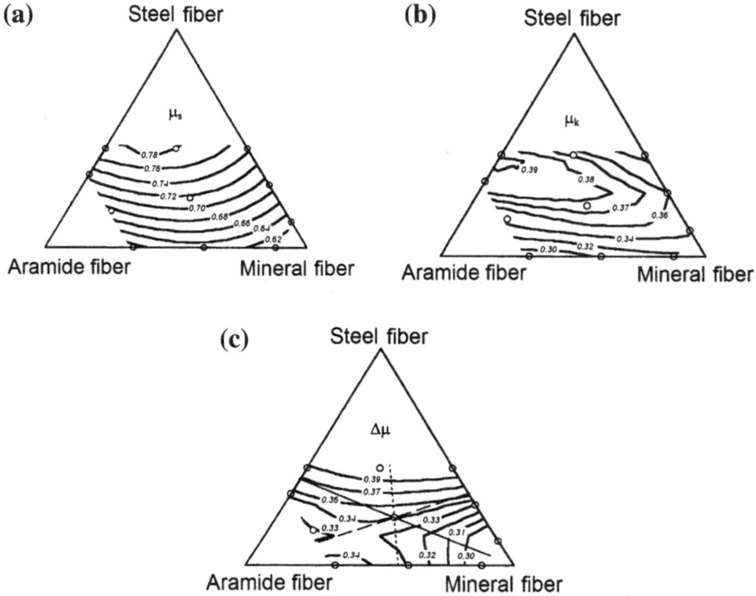


**Fig. 7.6** Effect of fibrous fillers on static (**a**) and kinetic (**b**) friction coefficients, and their difference (**c**)

### *7.3.1 The Effect of Fibrous Fillers*

Introduction of fibrous fillers in the friction composite formula depends, first of all, on the purpose to raise mechanical strength of the friction items. Besides, the fibrous fillers exert a direct effect on physico-chemical processes during rubbing and are able to alter friction characteristics of the composite. Although a great number of fibrous fillers is used presently in manufacture of frictional materials, the test results point to three types that differ much in their properties (heat conduction, mechanical strength, heat stability and adhesive energy) and are very indicative in this respect. They are the steel, aramide and mineral fillers [10]. The effect of the fibrous fillers on staticokinetic characteristics of friction was studied on 11 samples prepared with different fibrous filler content in the range similar to the volume concentration of the design shown in Fig. 7.5c. The measurement results of the static $\mu_s$ and kinetic $\mu_k$ coefficients of friction, as well as the difference between them $\Delta\mu = \mu_s - \mu_k$ are shown in Fig. 7.6. The experimental data were approximated by the second-order polynomial with determination coefficients ($R$-square) 0.97, 0.87 and 0.85 as shown in Fig. 7.6a–c, correspondingly. The approximation results in the form of the contour diagrams (Fig. 7.6a, b) have proved that a simultaneous increase in the static and kinetic friction coefficients is observed when the steel fiber content increases and that of the mineral fibers decreases. The filling by aramide fibers does not exert any essential effect in contrast to steel and mineral ones. From the other hand, filling by steel fibers raises, while by the mineral lowers the difference between the static and kinetic friction coefficients, which is significant in varying the friction-excited self-vibration level (Fig. 7.6c). It follows, that the optimal volume concentration is found for each component from the minimum condition $\Delta\mu$. For the case discussed in [10] this value amounts to (vol.%) 4:4:15, for the aramide, steel and mineral fibers, respectively.

## 7.3.2 The Effect of Matrix and Organic Fillers

The properties of the matrix polymers and organic fillers employed in friction composites can hardly be forecasted because of their inner thermal instability. Most applicable in manufacture of the polymer frictional materials are the PFR-based binders of the novolak or resol types. In addition to above named materials, synthetic rubber and cashew nut powder are frequently used to modify the matrix phase. Besides, filling by the cashew powder reduces the wear rate of the material at low temperatures, and augments stability of the friction coefficient at elevated temperatures.

The approximation results of the experimental data are illustrated in the contour diagrams (Fig. 7.7) where the determination factors are the next: 0.79, 0.81 and

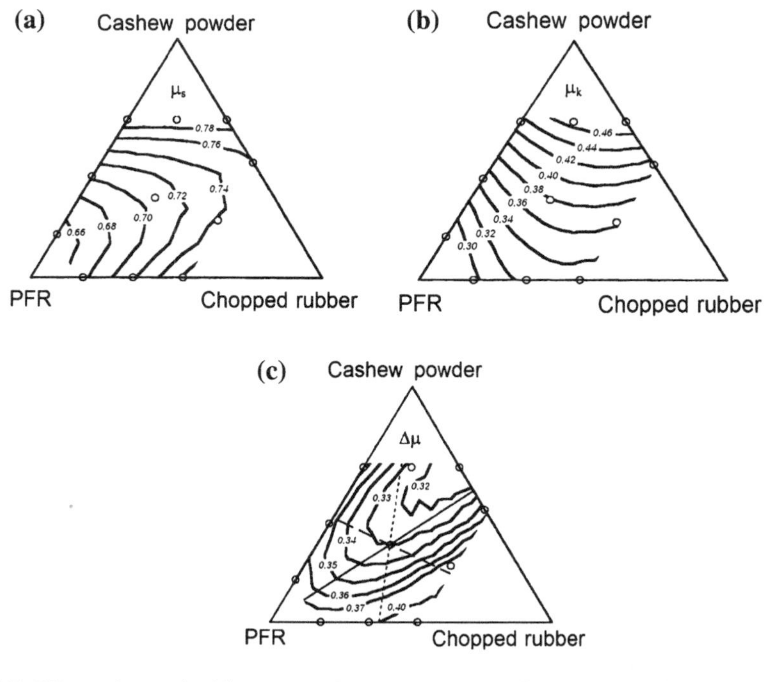


**Fig. 7.7** Effect of organic fillers on static (**a**) and kinetic (**b**) friction coefficients and their difference (**c**)

0.89 for Fig. 7.7a–c, correspondingly. It has been proved that as the PFR concentration increases the static friction coefficient drops. The kinetic friction coefficient increases if to raise the cashew powder concentration or to lower that of the PFR. Thus, the incremented PFR content brings about undesirable growth of $\Delta\mu$ From the other hand, the increasing cashew content effects favorably the amplitude of the friction-excited self-vibrations since the $\Delta\mu$ value drops in this case (Fig. 7.7c). The optimal ratio of organic fillers is 5:14:9, correspondingly for the PFR, cashew powder and rubber.

### *7.3.3 The Effect of Friction Modifiers*

In general, we may differentiate between two groups of friction modifiers. Firstly, these are greases and abrasive particles able to establish a compromise between the required friction coefficient value, wear resistance, and hostility towards the counterbody material. The abrasive particles are introduced into the fiction composites to monitor the friction coefficient, as well as to remove the transfer films formed on the rubbing surfaces as a result of pyrolysis. The aim of doping with greases is on the main to hinder wear. In the real formulas of friction materials more often a combination of abrasive particles and solid lubricants is used in order to provide for a needed friction coefficient value and formation of a lubricating layer within a wide temperature interval.

The dependence of both static and kinetic friction coefficients as well as their difference upon the volume concentration of different dispersed fillers (e.g., $ZrSiO_4$, $Sb_2S_3$ and graphite) is described in Fig. 7.8. The derived experimental evidences have been approximated by the second order polynomial with coefficients of determination 0.87, 0.91 and 0.90 as shown in Fig. 7.8a–c, correspondingly.

The increasing volume content of $ZrSiO_4$ particles or respective reduction of $Sb_2S_3$, concentration leads to a considerable increment in both static and kinetic coefficients of friction. This results in determination of the minimal $\Delta\mu$ value under the following volume ratios 11.0:3.5:1.5 of graphite, $Sb_2S_3$ and $ZrSiO_4$, correspondingly.

### *7.3.4 Optimization Results*

Optimization of the friction composite formulas [10] aims at provision of a minimal difference between the static and dynamic friction coefficients in all three groups of ingredients. It should be, however, noted that the minimal $\Delta\mu$ value is not a single criterion of a frictional material quality even in respect to vibroacoustic properties. It is important in the development of the real compositions to ensure a required level and stability of a number of other characteristics in a wide temperature range. These are the friction coefficient value, wear resistance, shear strength, susceptibility to

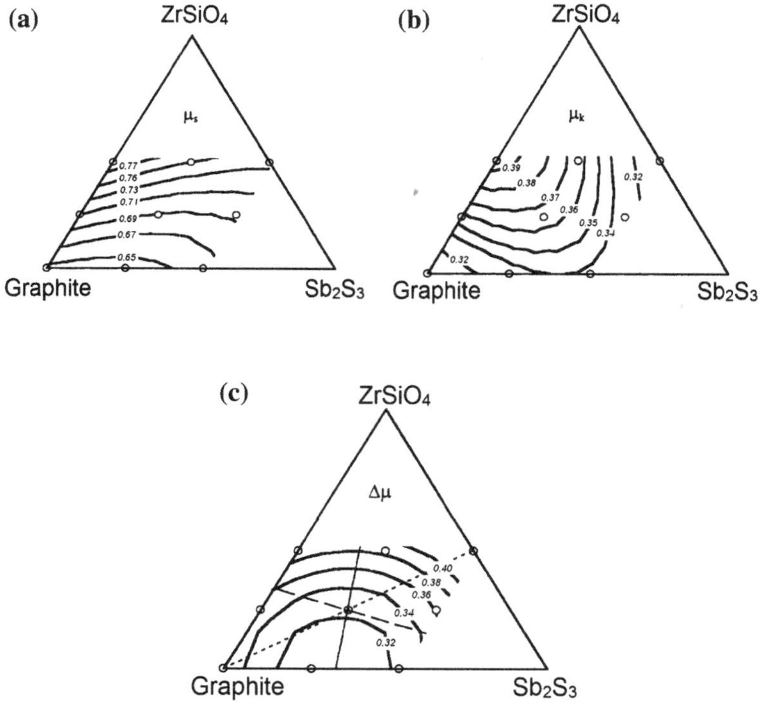


**Fig. 7.8** Effect of friction modifiers on static (**a**), kinetic (**b**) friction coefficients and difference between them (**c**)

generate HF noise (estimated as damping capacity) and forced vibrations (compressibility, aggressiveness towards the counterbody material, thermophysical properties and etc.).

For the case in question, the optimal concentration of the fillers has been estimated for all three groups proceeding from the minimum $\Delta\mu$ under variable $\mu_k$ within admissible variation range ±10 %. The rest of above-named quality criteria were not considered in the work [10]. Notice that the kinetic friction coefficient value of the modified samples remained within the admissible interval (±10 %) while concentrations of the fibrous fillers or friction modifiers were varied within the chosen range. The results in the group of organic fillers have shown that the choice of the composition with the minimal $\Delta\mu$ does not promote $\mu_k$ variation in the acceptable range. The modified composition selected as an optimal one with account of the corresponding limitations is presented in Table 7.1. The comparative experimental data for the initial and modified materials with the evident scatter of the results are shown in Fig. 7.9. The data prove that the difference between the static and kinetic friction coefficients can be minimized essentially by way of reducing the latter one.

The kinetic characteristic of friction (kinetic friction coefficient versus sliding velocity) is illustrated in Fig. 7.10 for the initial and modified materials under

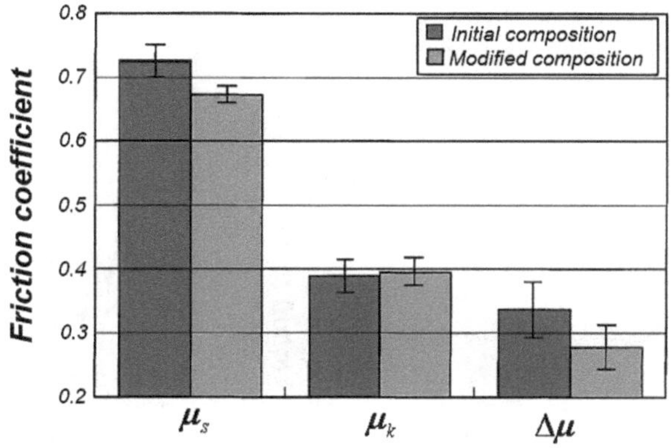


**Fig. 7.9** Static ($\mu_s$), kinetic ($\mu_k$) friction coefficients and their difference ($\Delta\mu = \mu_s - \mu_k$) for initial and modified compositions

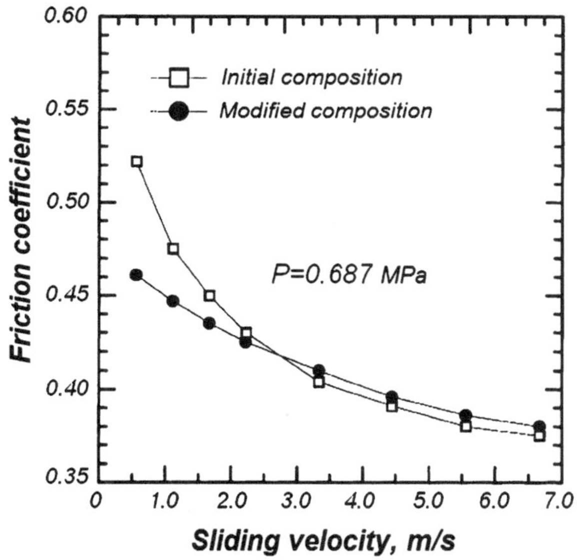


**Fig. 7.10** The kinetic frictional characteristic for initial and modified formulas of friction materials

0.687 MPa pressure. The data from Fig. 7.10 show that the curve characterizing the formula with the least difference between the static and kinetic friction coefficients has acquired a less expressed negative gradient $\mu_k$ of the sliding velocity (the curve is less sloping), which promotes abatement of the friction-induced self-vibrations (see Chap. 4). The friction force dependence in time for four constant sliding velocity values (in a stationary friction mode) is shown in Fig. 7.11. Since the disc thickness variations were negligibly small, we may conclude that the oscillating character of the friction force variation in time support the presence of the friction-induced self-vibrations in the brake system.

Figure 7.12 presents the amplitude of the friction force fluctuations as a function of sliding velocity for the materials with the initial and modified formula. The figure shows that the modified formula suppresses the fluctuation amplitude of the friction

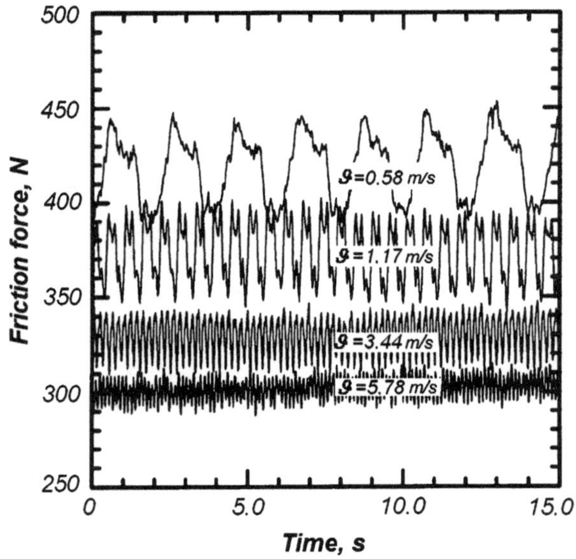


**Fig. 7.11** Friction force fluctuations at constant sliding velocities

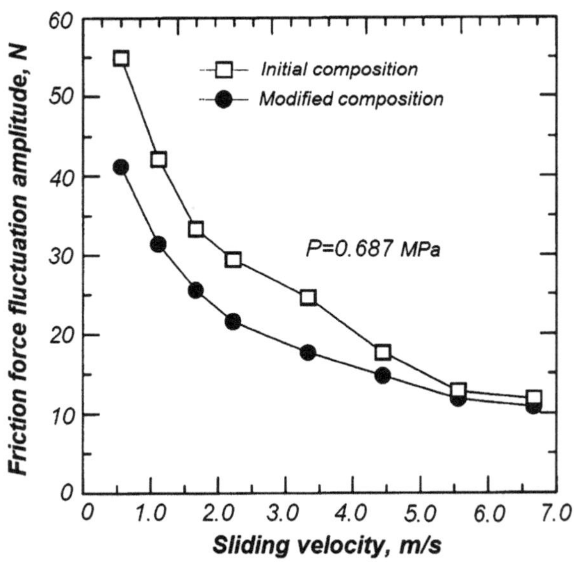


**Fig. 7.12** Dependence of friction force fluctuation amplitude on sliding velocity for initial and modified formulas of friction materials

force, especially at a low sliding velocity (till 20 %). This, most probably, lowers the vibration and noise loading, which, however, is not supported in [10] by the quantitative estimates of the parameters directly characterizing the very vibration and noise level generated by the brake unit.

Analogous results were obtained by the authors of [38–41] during vibroacoustic experimental investigations of multidisc oil-cooled (wet) brakes (MDOB) of the heavy-duty trucks. The investigations have proved that the traditional by their

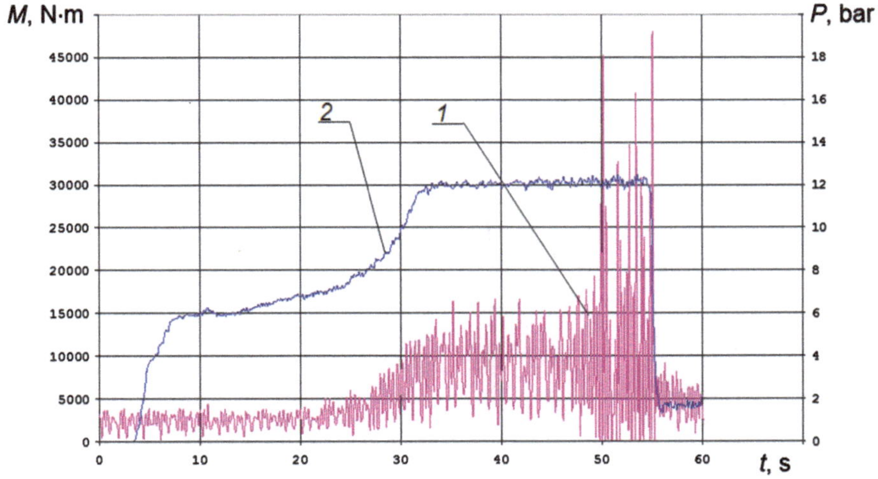


**Fig. 7.13** Kinetics of friction torque variation (*1*) and pressure (*2*) variations in MDOB drive in FM tests with the original structure

composition and structure polymer friction materials used in a new design of the MDOB promote a significant growth of the amplitude of the LF oscillations of the brake torque and vibrations of the brake system in compliance with Fig. 7.13.

The bench tests of the brake discs (general view of the bench is presented in Chap. 5, Fig. 5.8) have confirmed the conclusion that stability of the brake torque is interrelated with the MDOB vibration level and a relative slippage of the friction discs. The moment of slippage corresponds to the maximal normal to the contact surface acceleration of the friction discs relative to each other and an abrupt friction torque reduction. It should be underlined that the level of friction-excited self-vibrations of the MDOB in this situation depends much upon the composition and structure of the friction layer material of the brake discs.

Special materials with enhanced properties may assist in eliminating above-named drawbacks, e.g., friction materials with a high-porous polymeric matrix. The main task in this respect is to create a highly strong heat-proof polymeric structure in the FM able to maintain a liquid or a boundary lubrication regime of the rubbing solids with the least difference between the static and dynamic friction coefficients independently of the lubricant viscosity. The newly developed porous anisotropic FM discussed in work [42] are said to be able to lower the probability of the oil film tearing off the contact surface and transfer to a dry friction. The tests of the FM with optimal for the given service conditions structure have proved that the amplitude of the friction torque oscillations and vibration in the brake unit are decreasing considerably (Fig. 7.14).

If to proceed from the chosen approach on the base of mathematical methods of optimizing multicomponent systems and experimental design, it will be impossible to explain how the components of the friction composite effect instability of the friction process, in particular, characteristics of self-oscillations. Evidently, to

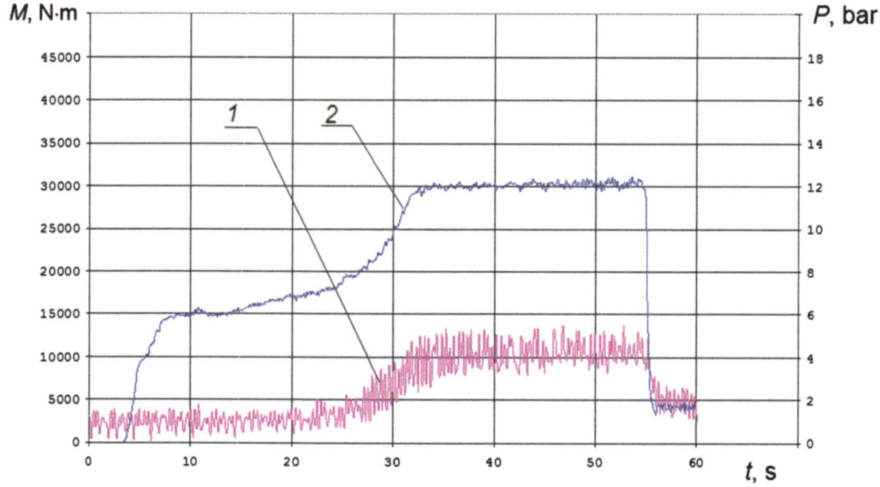


**Fig. 7.14** Kinetics of friction torque variation (*1*) and pressure (*2*) variations in MDOB drive in FM tests with the modified structure

understand this mechanism we should perform a detailed investigation on how each ingredient behaves in the considered processes with the help of modern techniques like the atomic force microscopy (AFM), scanning electron microscopy (SEM) and other.

## 7.4 Optimization of Composition and Dynamic Mechanical Characteristics of Friction Materials

The investigation results presented in [43–45] disclose the effect of the formula and structure of the FM used in car brake and transmission designs on their dynamic characteristics and HF vibroacoustic activity of the friction joints. The objects of investigations were high-filled friction composite materials based on thermosetting binders (liquid and powder phenolic resins of the resol and novolak types). The novolak resins were cured by hexamethylene tetramine (8 mass%). The main (by the mass) dispersed and reinforcing fillers were metal oxides, barites, basalt, glass, carbon and lingo-cellulose fibers; steel and brass shaving and copper powder. Viscoelastic properties were monitored by modification of the binder as well as dressing of the fibrous and metallic fillers using structural modifiers like butadiene-nitrile rubber with the mean content of acrylonitrile chains 28 %, cashew shell liquid, polyvinyl acetate and a mixture (in equal proportion) of triglycerides of palmitic $C_3H_5(OOCC_{15}H_{31})_3$, linolic $C_3H_5(OOCC_{17}H_{31})_3$ and linoleic $C_3H_5(OOCC_{17}H_{29})_3$ acids. The cashew nut shell liquid (CNSL) presents a mixture of the mono and diatomic alkyl phenols of cardanol and cardole in the ratio 9:1. After thermal treatment the molecules of these compounds include an unsaturated aliphatic substitute of the averaged composition $R^{(1-3)} = C_{15}H_{27}$ when the phenolic ring is found in the meta-position.

The test samples were made by thermal compression under (458 ± 5) K temperature. The friction compositions differed in their formula, the filler and binder content, as well as in technological parameters of forming. The relative degree of hardness of the FM based on a thermosetting polymeric matrix was found by the soluble matter content extracted from the specimen material. The Brinnell hardness was measured according to GOST (State Standard) 9012–59 using a durometer TR 5006 M (GOST 23677–79). Hardness of the samples was 16–49 HB. To make vibroacoustic testing on a full-scale test bench the FM samples were chosen with the friction coefficient within 0.55 ± 0.12 and different dynamic characteristics. Mentioned friction coefficient levels are in line with the requirements to frictional efficiency of the real friction joints. For comparison the samples of sintered cermet FM on the brass-bronze base were used.

### *7.4.1 Tribological Tests*

The friction tests were carried out on lab tribometers SMT-1 and I-32 M-1 following the geometries like rotating disc—immovable indenter and Vee-block (partial insert) on ring under stationary friction modes. The test schemes are illustrated in Fig. 7.15.

The pressure in the friction joint was varied within 0.5–2.5 MPa; and the linear velocity within 0.5–2.5 m/s. The counterbody material was carbon steel 65G (GOST 14959-79) of $HRC_э$ 35–37 hardness and surface roughness $R_a \leq 1.25$ μm.

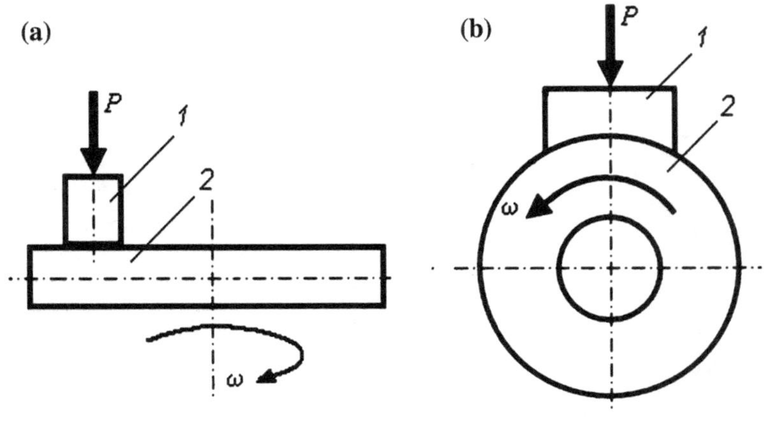


**Fig. 7.15** Friction geometries: **a**—rotating disc—immovable indenter; **b**—Vee-block—ring; *1*—FM sample, *2*—counterbody

The tests were conducted under (295 ± 2) K temperature in air at dry friction. The full-scale inertial-brake stands simulating the real operating regimes of the brakes and transmissions were used to study the real friction joints in nonstationary friction conditions.

### 7.4.2 Determination of Dynamic Characteristics of Materials

The dynamic characteristics of materials, namely, the dynamic elasticity modulus and coefficient of mechanical losses were studied using the dynamic-mechanical analysis, including the non-resonant method to define temperature dependencies, and the method of the resonant amplitudes to find the dependencies on the static load.

*The non-resonant method* involves the amplitude and phase shift measurements of the signals of the driving force and resultant deformation when the sample material experiences forced harmonic vibrations at the frequencies much less the resonant ones. This method makes possible to determine dynamic characteristics of materials as a function of time, temperature and frequency.

A dynamic mechanical analyzer DMA-8000 (Perkin Elmer) was used for measurements under up to 293…673 K temperatures. A scheme of the measurement path of the dynamic mechanical analyzer is shown in Fig. 7.16.

*The method of resonant amplitudes* is based on determination of the dynamic model parameters of a viscoelastic body (Voight-Kelvin's model) at longitudinal vibrations of the sample (Fig. 7.17a).

This method consists in finding resonant frequency $f_r$, under which the vibration amplitude and the coefficient of resonant amplification $A$ increase abruptly, where $A$ equals to the ratio of dynamic displacements under the resonance to induced by the

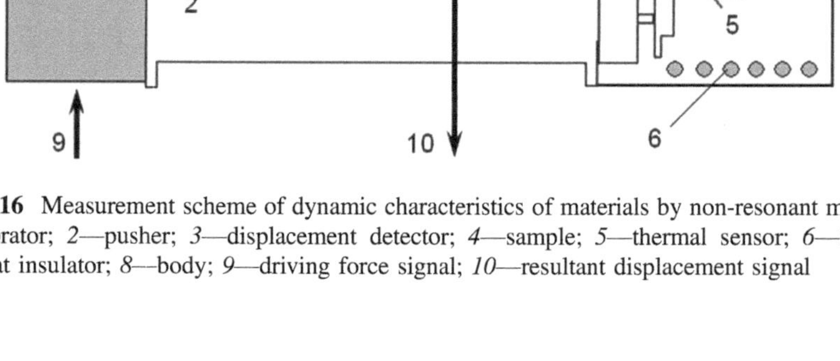


**Fig. 7.16** Measurement scheme of dynamic characteristics of materials by non-resonant method; *1*—vibrator; *2*—pusher; *3*—displacement detector; *4*—sample; *5*—thermal sensor; *6*—heater; *7*—heat insulator; *8*—body; *9*—driving force signal; *10*—resultant displacement signal

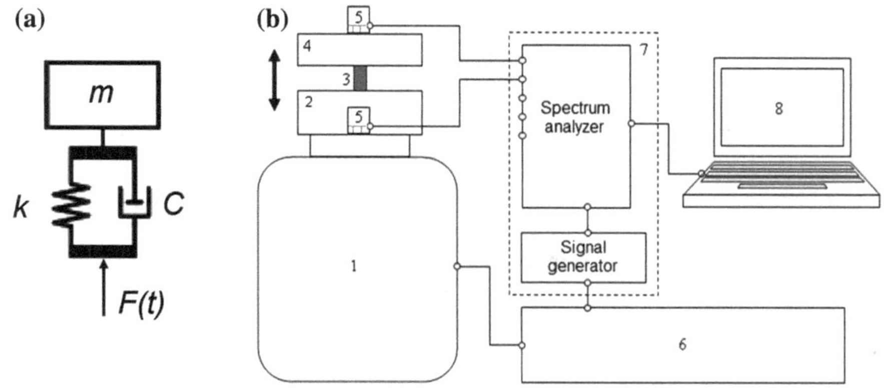


**Fig. 7.17** Determination of dynamic characteristics of materials by the resonant amplitude method: oscillating system with one degree of freedom based on the Voight-Kelvin's model for estimating viscoelastic behavior of materials **a** scheme of measurement device **b** *1*—shaker; *2*—shaker table; *3*—sample; *4*—weight; *5*—accelerometer; *6*—amplifier, *7*—data acquisition and analysis system

static load displacement. The scheme of the measurement device is shown in Fig. 7.17b. The test system includes an electrodynamic vibrostand S522 fit with accelerometers 4513-001 (Bruel and Kjaer), signal generator and a PC-aided recording system Pulse 3560B (Bruel and Kjaer).

The dynamic elasticity modulus $E_d$ of the material was found according to the measurement scheme by the formula:

$$E_Д = \frac{4\pi^2 f_p^2 mh}{S},$$

where *m*—weight mass, kg; *h*—sample height under loading, m; *S*—area of the sample, $m^2$.

The loss factor η was found from the equation

$$\eta = \frac{1}{\sqrt{A^2 - 1}}.$$

It has been proposed in Chap. 4 of [46] to use a reduced index as a verification parameter characterizing damping capacity of the *i*-th FM modification with account of the elementary oscillators calculated according to the formula

$$\overline{D}_i = \frac{(E_Д \cdot \eta)_i}{(E_Д \cdot \eta)_{\max}},$$

where $(E_Д \cdot \eta)_i$ and $(E_Д \cdot \eta)_{max}$ are a product of the dynamic elasticity modulus by the loss factor measured by the resonant method under 20 ± 2 °C temperature and 0.25 MPa static load for the *i*-th material and a sample with the best damping capacity (for the samples under study).

### 7.4.3 Noise and Vibration Measurement Procedure for Friction Joints

It seems justifiable to unite the friction and vibroacoustic bench tests [13, 14] into one approach in order to optimize the friction joint materials in respect to their vibroacoustic characteristics. Acoustic measurements in special tight chambers are considered to be a requisite condition in achieving trustworthy experimental evidences about noise. The major problem of the combined tests in studying susceptibility of frictional composites to generate noise and vibration in the rooms not conditioned and rigged specifically for vibroacoustic measurements consists in the difficulty or even improbability to recognize the features resulted from the tribological behavior and the properties of the FM used in the friction joints in the whole noise spectrum. The problem of identification of the frictional components of noise is solved in the given work by the coherent analysis of the noise and vibration signals measured simultaneously. The laser Doppler vibrometer makes possible to perform the contactless measurements of vibration in any accessible sites of a system, including high-temperature rotating elements of the friction joints. The measurements of the level and intensity of acoustic radiation generated by the friction pair were conducted by a double-microphone intensimetric probe of a narrow directional pattern. The contactless measurements directly in a tribopair have allowed us to abate the effect of external vibrations transmitted over the body and conjugated parts and that of the background noises always present in a test lab. So, the components not connected with the friction processes in the tribopair under study were excluded from the spectra. Thus, the reliable experimental data were obtained based on a high selectiveness and informativity of the technical means employed. They assisted in identification of the noise and vibration frictional components, and achieving an adequate estimate of the level and probability of their generation in usual conditions without application of a special anechoic chamber.

Noise and vibration measurements in tribopairs were conducted under both stationary and nonstationary regimes by a laser Doppler vibrometer VH-1000D (Ometron) and an acoustic intensimeter 3599 (Bruel & Kjaer). This was combined with signal processing by the computer and numeric analyses. The general schemes of the test bench and accompanying technical facilities for noise and vibration measurements in tribopairs are shown in Fig. 7.18.

Evaluation characteristics reflecting the efficiency of a structurally modified FM usage in a friction joint in respect to its vibroacoustic activity include the relative noise frequency $W(N)$ and its level reduction ($R$) as a function of the damping capacity index of the material. The relative noise frequency is presented by a ratio of the test number during which a noise is generated at the level suffice for its objective identification by the methods employed to the total number of actually performed tests. The noise level reduction in the friction joint incorporating a structurally modified FM was found by the formula:

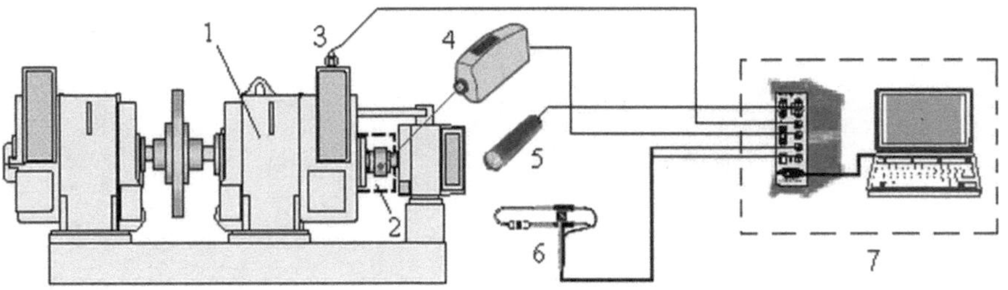


**Fig. 7.18** Diagram of an experimental setup: *1*—test bench, *2*—friction joint, *3*—triaxial accelerometer, *4*—laser Doppler vibrometer, *5*—strain gage, *6*—intensimetric probe, *7*—data storage and processing system

$$R_i = 20\lg(p_i/p_{\max}),$$

where $p_i$ is the mean acoustic pressure generated during tests of the *i*-th material, $p_{\max}$—mean acoustic pressure at testing materials generating the highest noise level.

### *7.4.4 Structure and Composition Effect on Dynamic Mechanical Characteristics of Frictional Materials*

The dynamic characteristics of FM have been studied experimentally as a function of the polymer matrix type, employed plasticizers of the matrix phase with various formulas and chemical structures, hardening degree of the binder, as well as geometry and orientation of the fibrous fillers.

*Modification of the matrix phase*. Table 7.3 presents data on the damping capacity of the FM with different matrix types and plasticizer grades.

**Table 7.3** Damping capacity of FM with different polymeric matrices

| Matrix | | Description | $\overline{D}$ |
|---|---|---|---|
| Type | I | Phenolic resin of resol type | 0.07 |
| | II | Phenolic resin of novolak type | 0.08 |
| | III | Mixture of resol and novolak phenolic resins | 0.09 |
| | IV | Rubber-polymer matrix[b] | 1.00 |
| | V | FM with cermet matrix[a] | 0.17 |
| Plasticizers | VI | Rubber (5 mass. %)[a] | 0.31 |
| | VII | Rubber (15 mass. %) | 0.35 |
| | VIII | Mixture of triglycerides of fatty acids | 0.23 |
| | IX | Polyvinyl acetate | 0.26 |
| | X | CNSL | 0.32 |

[a] test materials in friction joints under study showed the highest noise level; their mean values were used as $p_{max}$ to calculate $R_i$

[b] damping capacity chosen as the maximal to calculate $\overline{D}_i$

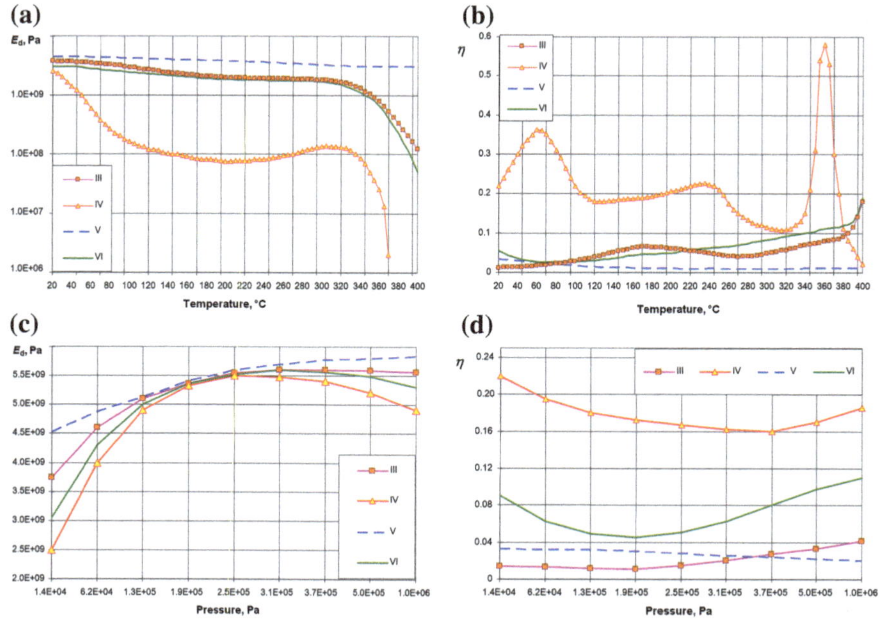


**Fig. 7.19** Dynamic elasticity modulus (**a**, **c**) and loss factor (**b**, **d**) versus temperature (**a**, **b**) and load (**c**, **d**) for FM with different matrix types

Figure 7.19a–d illustrate the dependencies of the dynamic properties of some model frictional composites upon temperature (Fig. 7.19a, b) obtained by the non-resonant method, and static compressing load (Fig. 7.19c, d) found by the method of resonant amplitudes.

According to Table 7.3 and Fig. 7.22, the FM with a non-modified matrix (matrices I, II and III) display the lowest damping capacity. The damping capacity value for these samples turned to be somewhat lower than the one for the cermet FM. However, the damping capacity of the polymer-based FM rises essentially with the temperature growth for all studied materials and may vary within a wide range depending on the composition and structure (see Fig. 7.19a–d). The dynamic characteristics show mot intensive rise when a rubber-polymer matrix is used. Modification of the initial FM (matrix I) by structural modifiers (matrices VI–X). also exerts a noticeable effect

*Effect of cure degree.* Figure 7.20 illustrates the temperature dependencies of the polymer matrix-based FM properties with different degrees of cure ($c$). The degree of cure is most pronounced in the materials based on the rubber-polymer matrix. The experimental results on different types of polymeric matrices show that the increasing degree of cure generally leads to elevation of $\overline{D}$ value mainly because of the $E_{\text{Д}}$ growth. The dissipation component value $\eta$ is in fact independent of the cure degree (see Fig. 7.20) but is conditioned directly by the FM structure and formula, and by its matrix type on the first place.

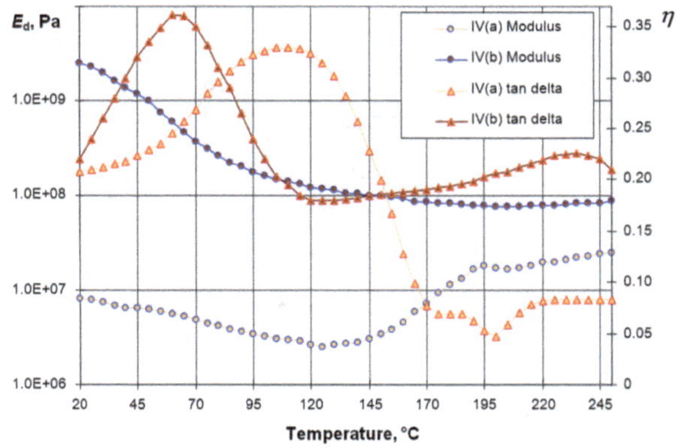


**Fig. 7.20** Dynamic elasticity modulus and loss factor versus temperature for the rubber-polymer matrix-based FM (matrix IV) with different cure degree $c$: 0.50 (**a**); 0.85 (**b**)

**Table 7.4** Damping capacity of FM with different fibrous fillers and their orientation

| Filler | Geometry | | Matrix | Orientation* | $\overline{D}$ |
|---|---|---|---|---|---|
| | Diameter, μm | Length, mm | | | |
| Basalt | 0.6–3.0 | 1.0–1.5 | I | = | 0.03 |
| | | | | ⊥ | 0.03 |
| | | | IV | = | 0.91 |
| | | | | ⊥ | 0.90 |
| Glass roving (6 filaments in a roving) | 13 | 5–7 | I | = | 0.11 |
| | | | | ⊥ | 0.07 |
| | | | IV | = | 1.00 |
| | | | | ⊥ | 0.90 |
| Glass fiber | 6–10 | 15–20 | I | = | 0.14 |
| | | | | ⊥ | 0.05 |
| | | | IV | = | 0.99 |
| | | | | ⊥ | 0.89 |
| Lignocellulose | 800–2,000 | 2.0–5.0 | I | = | 0.15 |
| | | | | ⊥ | 0.14 |
| | | | IV | = | 0.96 |
| | | | | ⊥ | 0.95 |

* *Note* Symbols «=» and «⊥» stand for, correspondingly, longitudinal and transverse directions of the dynamic force relative to reinforcing fibers in the FM samples

*Effect of geometry and fibrous filler orientation.* Table 7.4 presents experimental results on damping capacity of the materials with matrices I and IV reinforced by the fibrous fillers imparted different geometries and oriented in two directions (parallel and perpendicularly) relative to the dynamic force action on the samples as according to Fig. 7.21.

The dynamic tests by the method of resonant amplitudes have proved that filling by the fibers and their orientation may result in a number of cases in improved

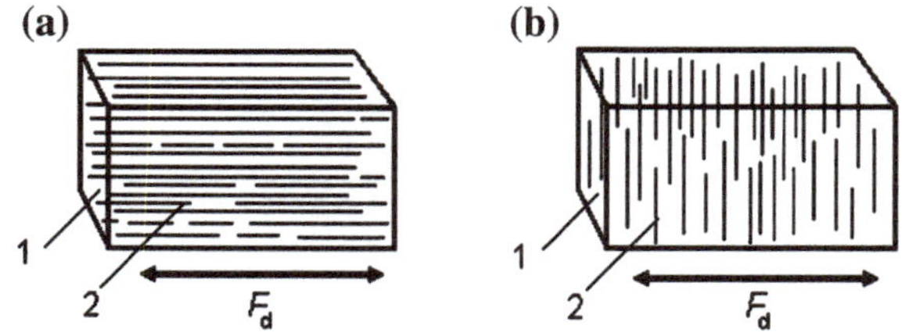


**Fig. 7.21** Orientation of fibrous fillers in the FM sample relative to dynamic force action $F_d$: **a**—parallel, **b**—perpendicular; *1*—matrix, *2*—fibrous filler

damping capacity of the materials mainly via the $E_д$ growth. Named factors do not exert any noticeable effect on the loss factor except for the lignocellulose particles. The results presented in Table 7.4 confirm that $\overline{D}$ increases in the next cases:

- the matrix is reinforced by the long fibers;
- the fillers with a fibrous-porous structure are used, e.g., lignocellulose particles;
- the FM has acquired a special structural anisotropy (reinforcing fibers are directed along the dynamic force action).

Mentioned regularities are preserved in various types of matrices studied in the present work.

Figure 7.22 illustrates the dependencies of the absolute damping capacity value upon temperature (Fig. 7.22a), and those for the static compressive load (Fig. 7.22b) based on the measurement results of the dynamic mechanical characteristics of the FM with a metallic matrix (curve 2), and PCFM (curves 1, 3 and 4). The damping capacity of the PCFM shows a strong dependence on the temperature and varies in a wide range in response to the composition and structure. The FM with a metallic matrix display stronger stability of their parameters in response to temperature, and lower loss factor values that are decreasing with temperature rise as compared to the polymeric FM.

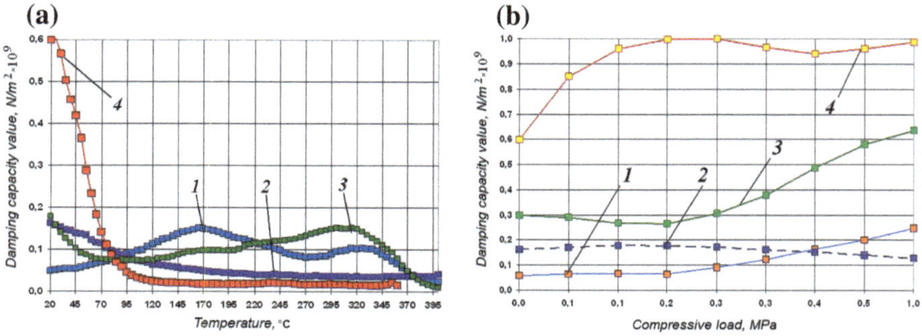


**Fig. 7.22** Damping capacity value of FM as a function of temperature (**a**) and static compressive load (**b**): *1*, *3*, *4*—PCFM; *2*—metal-ceramic friction composite on bronze base

### 7.4.5 Results of Tribological and Vibroacoustic Tests of Friction Materials

*Tests on a laboratory tribometer.* A series of materials that display high enough friction and wear characteristics along with desired friction efficiency have been selected for vibroacoustic and tribological tests using a laboratory tribometer and a full-scale brake dynamometer simulating the real operating modes of the real friction systems of machines. Table 7.5 lists characteristics of some FM having considerable differences in their damping capacity and insignificant deviations from the parameters for vibroacoustic activity discussed earlier.

The kinetic characteristics of the tribopair materials tested in a laboratory tribometer are presented in Fig. 7.23. The typical noise spectra generated by the friction joint at testing two FM with a polymer matrix having essentially different damping capacity are shown in Fig. 7.24. The friction components of the spectra used for

**Table 7.5** Characteristics of friction materials

| $i$ | Friction material (FM) | Friction coefficient | Hardness, HB, MPa | $I_h \times 10^{-8}$ ($P$ = 1.0 MPa; $V$ = 1 m/s) | $D \times 10^8$, H/m$^2$ | $L_p$, dB |
|---|---|---|---|---|---|---|
| 1 | FM non-modified by polymeric matrix (initial composition) | 0.44–0.49 | 46–49 | 2.62 | 0.84 | 82 |
| 2 | FM with metallic matrix | 0.30–0.35 | 36–41 | 2.0 | 1.57 | – |
| 3 | FM plasticized by polymeric matrix (modified composition) | 0.60–0.67 | 29–31 | 3.0 | 2.81 | 45 |
| 4 | FM based on rubber-polymer matrix (modified composition) | 0.49–0.54 | 16–19 | 2.9 | 9.19 | 33 |

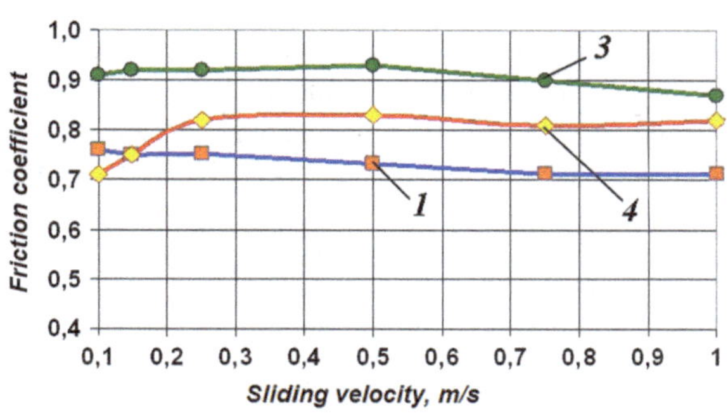


**Fig. 7.23** Friction coefficient versus sliding velocity for FM with a polymeric matrix

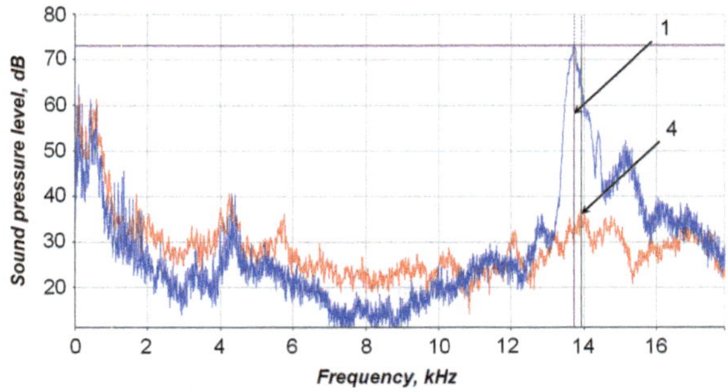


**Fig. 7.24** Noise spectra generated by a rubbing in the tribometer tribopair made of FM with different damping capacity values ($P$ = 0.25 MPa, $\vartheta$ = 0.15 m/s; $T_1$ = 87 °C; $T_2$ = 55 °C; $\mu_1 = \mu_2$ = 0.75)

vibroacoustic estimates of the friction pair ($f_1$ = 13.5 kHz; $f_2$ = 13.9 kHz) are indicated by the arrows.

The results of the composition optimization are analyzed on the examples of materials 1, 3 and 4 according to Table 7.5. The material with the highest kinetic friction coefficient within the whole studied load and velocity range (curve 3 in Fig. 7.23) shows a lower noise level (Table 7.5) as compared to the material having a less kinetic friction coefficient (curve 1 in Fig. 7.23). This can be attributed, firstly, to considerable differences in damping capacity values of the materials (it is thrice as high in material 3), and secondly, to a small positive gradient of the kinetic friction characteristic (Fig. 7.23). The kinetic friction characteristic in the low-velocity range (till 0.25 m/s) is explicitly growing only for material 4 (Fig. 7.23, curve 4). This, together with a much higher damping capacity ensures the friction-induced noise component reduction by more than 30 dB. It should be emphasized that such a significant effect like that is observed only in the case of a stable temperature on the friction surface corresponding to a maximal one (or close to it) of the material damping capacity (Fig. 7.22a) and a positive friction coefficient gradient relative to a sliding velocity (Fig. 7.23). In the above-considered case ($P$ = 0.25 MPa, $\vartheta$ = 0.15 m/s), the friction surface temperature was within 55… 87 °C, which corresponded to the region of elevated (in contrast to the initial one) damping capacity of the materials and to a positive friction coefficient gradient for the sliding velocity of materials 3 and 4 with a modified formula.

*Dynamometer tests*. The friction effectiveness, smoothness of engagement and vibroacoustic activity of clutches have been determined on a full-scale drag dynamometer simulating the real operating modes of the wheel tractor transmission. A general view of the dynamometer is shown in Fig. 7.25 [46].

Figure 7.26 presents investigation data on the frequency and levels of noise origination obtained in the laboratory and frictional bench tests of the friction discs lined with the materials (according to the table) having a close friction coefficient and hardness but different damping capacity values The amplitude-frequency characteristics of vibration and noise generated by similar FM used in different

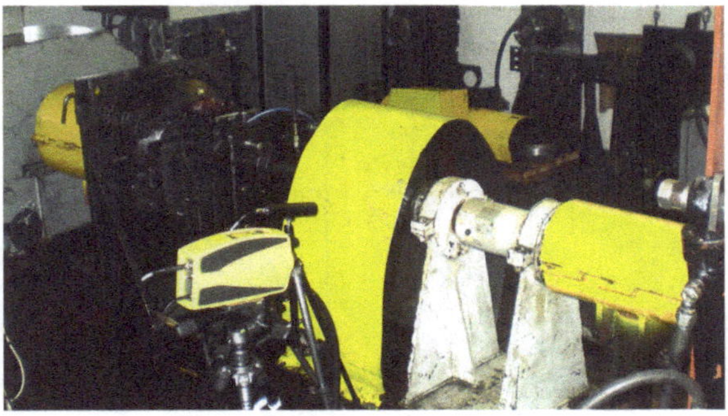

**Fig. 7.25** A general view of a full-scale drag dynamometer for testing clutch discs [46]

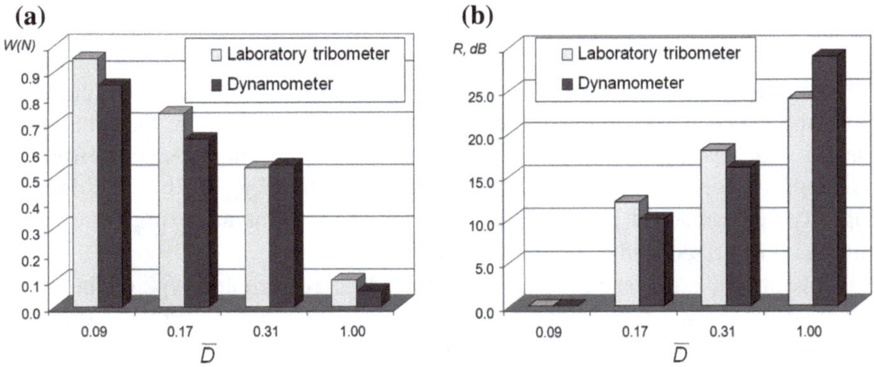


**Fig. 7.26** Effect of dynamic characteristics of FM on the relative frequency (**a**) and reduction of noise level origination in conditions of stable friction-excited self-vibrations (**b**) at material testing in different tribojoint designs: *1*—laboratory tribometer; *2*—full-scale drag dynamometer

tribopairs have displayed considerable differences, which is supported by their structural dynamics. It is to be noted that the regularities of the damping capacity effect of the studied FM upon the relative vibroacoustic frequency and level of tribojoints established on a lab tribometer agree well with those observed in the bench tests (see Fig. 7.26). Proceeding from above evidences, we may conclude that the components of the friction-induced vibroacoustic spectra are in their essence the characteristics of the rubbing material structure and properties.

Figure 7.27 presents typical time dependencies of the acoustic pressure spectra appearing in a friction joint at nonstationary sliding. They are based on the test results of the FM with different damping capacity values (materials 2 and 4 according to Table 7.5).

The data presented in Fig. 7.27 visualize that rubbing of the discs lined with a cermet FM of a low damping capacity is accompanied by the intensive noise generation (till 105 dB) with an expressed frictional component on the 292 Hz

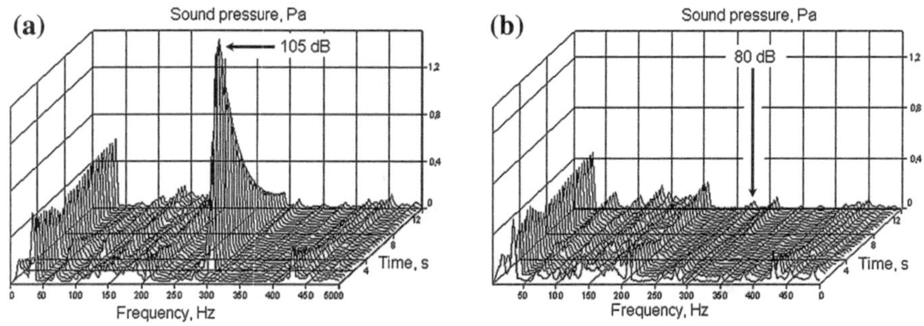


**Fig. 7.27** 3D FFT spectra of noise generated in the dynamometer during testing FM with different damping ability of cermet (**a**) and rubber-polymer modified matrices (**b**)

frequency. The mean noise level generated by this FM was used as a $p_{\max}$ in $R_i$ calculations.

It follows from Fig. 7.26 that the FM with a raised damping capacity used in the tribojoint decreases significantly the relative frequency and level of the noise generation. The acoustic pressure level induced by vibroactivity of the tribojoint in the fill-scale testing of the FM with the best dynamic characteristics (from the developed model FM samples) did not exceed 80 dB (Fig. 7.27b). The damping capacity value reached by the given material was used as a maximal one in calculations of the reduced $\overline{D_i}$.

Thus, it was established that the composition, structure and type of the matrix phase of the CFM predetermine their dynamic characteristics and their affinity to vibration and noise generation in friction joints. A procedure is proposed for determining noise and vibration levels and their probability estimation in tribojoints depending on the dynamic characteristics of the FM used in the tribojoint.

We believe that damping capacity plays one of the key roles in the HF vibroacoustic activity of the friction pairs. However, like in the case with the staticokinetic characteristics, it turns hard to direct optimization towards damping capacity increase within the maximum wide temperature interval because of the need to leave all other service characteristics of the material intact.

## 7.5 Methods of Forced Vibration Abatement in Brakes

Modern approaches to fighting hot vibration proceed from the positions of eliminating nonuniform heat generation in combination with averting propagation of the hot spots by making heat spreading over the friction surface more uniform. There exist certain means that play a significant role in decreasing vibration, namely: increase of the disc heat conduction and reduction of the heat expansion value in the tribopair materials; optimization of compressibility of the FM, minimization of the lining contact area (along the arc), mounting of the disc so as to exclude its conical bending, refinement of the disc design to prevent tapering, minimization of non-flatness and butt beat [47].

A traditional method of fighting cold judder used by in fact all car manufacturers consists in decreasing disc run-out. It is, however, practically impossible to hamper beating completely. For instance, deviations from the tolerances on dimensions arising at manufacture or assembly may result in uneven thermal loads and wear. To decrease DTV, the manufacturers of supports may reduce sliding friction or change characteristics of the piston seal recoil, which, however, may vary within the car service life. So, it is impossible to eliminate run-out full, wherefore, the role of the FM in originating DTV should be reduced to a minimum. High effectiveness of the materials science methods in this problem is supported by an important factor that before any changes in frictional composites connected with excluding asbestos fibers from their formula, the problem of the forced cold judder did not seem so pressing.

From the other side, a friction material should be fit with a combination of characteristics that may limit uneven wear of the disc and the related DTV [48]. It is critical for the manufacturers of FM to balance the properties of the materials so as to reach a compromise between, e.g., corrosion prevention and inhomgeneity of the disc thickness, providing all other important functions, like removal of metal corrosion products remain invariable.

### *7.5.1 Minimization of Thermal Deformations*

Various methods are used in practice to minimize heat-induced deformations, especially in the metallic brake discs. Thermal stability of the brake disc geometry depends on the material quality, thermal pretreatment before machining and the very design of the disc. Different designs of the brake discs in which their resistance to tapering is accounted for are discussed in numerous works [49–52] (see Chap. 5). It is possible to improve resistance of a disc to heat deformations by using high-carbon materials, or removal of thermal strains during machining. The technological factors influencing shape variations of the disc are discussed in works [53, 54]. Among the factors that may afterwards cause variations in disc geometry are filtering and cooling of the melt, location of the air gap, state of the raw material for molding (conditions and storage time), the parameters of thermal treatment of the disc, grinding before and after annealing, as well as position in the oven.

We can enumerate the following critical thermophysical properties of the brake disc [50, 55]:

1. Specific heat, i.e., the capacity to accumulate heat energy. A considerable amount of heat is accumulated during the initial period of braking [50], therefore, it is a dominating factor in the short-term braking.
2. Heat dissipation turns to be an important factor at a prolonged braking (2–3 min), i.e., in a slow-down mode over a continuous grade. The processes of heat transfer

are also influencing heat restorability between stops [56]. The share of the convective heat exchange makes up 90 % of the whole heat energy in almost all braking conditions [50], due to which the heat radiation is neglected. The coefficient of convective heat transfer is proportional to the car speed in the 0.8 power [50].

3. Heat conductivity is the ability to redistribute the heat energy. The maximal temperature values during the prolonged low-intensity braking events depend mainly on the heat conductivity of the material. The least effect of the heat conductivity is observed in a short-term braking [57].
4. The coefficient of heat expansion is connected with the processes of the friction contact localization caused by thermal strains. It defines disposition of the disc towards formation of the hot spots and DTV. The temperature gradients may appear as a result of momentary DTV or uneven heat expansion of the materials.

The lightweight brake discs on the aluminum matrix base, especially the ones reinforced by silicon carbide, are more sensitive to temperatures as compared to the discs based on grey cast iron (about 450 °C) [57]. Since named materials display a low heat capacity, they can be accepted only for the cars with the total mass till 1,000 kg. The use of aluminum composites or pure aluminum with a composite coating aggravates the problem of forced vibrations because of a high heat expansion coefficient and a low heat capacity [51]. In this case, the high heat conductivity will exert a negligible effect on generation of the hot spots and DTV.

Nevertheless, there do exist the materials, in particular ceramic ones that are able to decrease forced vibration in brakes. As an example, we can name silicon carbide materials (C/SiC) reinforced by short fibers that have only recently been adopted in the sport car and railway vehicle production [58]. Forced vibration is reduced in this case due to, primarily, a low heat expansion coefficient and low wear. From the other hand, a low elasticity modulus promotes more even distribution of the contact pressure, thus hampering the heat-induced DTV and hot spots. High wear resistance of ceramic materials enables to use them in heavy-loaded brakes of the heavy-duty machinery, where the traditional cast iron discs are usually unacceptable because of thermal cracking. However, these materials are rather expensive, which restricts their application in the main segment of the motorcar branch.

### *7.5.2 Optimization of Friction Material Properties*

To develop a target friction material a designer should take into account more than 20 parameters. They include density, thermal stability, strength (tensile, compressive, bending and shear), manufacturability, ecological safety, inclination to groan and squeal generation, and etc. [59]. In respect to forced vibrations, the parameters like compressibility, hostility towards the counterbody material are more important than others. The friction coefficient, the coefficient of heat expansion, heat conductivity, corrosion resistance and porosity are the factors also affecting forced vibration to a greater or lesser extent.

In general, we can isolate two approaches used to optimize the composition of the FM that are acknowledged as the effective means in abating forced vibration.

The essence of the first approach consists in creation of the FM less hostile to the brake disc, which may reduce its disposition to developing inhomgeneity across thickness. This, so-called passive approach, is most spread among the Japanese manufacturers of FM. However, the intrinsic initial polythickness of the disc, which is in fact unremovable over its whole service life together with the obstacles in attaining high enough kinetic friction coefficient are the main serious drawbacks of the approach in question. Among other disadvantages of this method there are insufficient degree of corrosion debris removal and healing of geometrical imperfections (smoothening) of the friction surface due to uneven wear as a result of friction-induced self-vibrations.

Another approach in contrast to above-described one consists in creation of the FM with improved abrasive properties aimed at elimination of DTV arising in conditions of normal braking from, e.g., periodic wobbling of the shoe disc in the case of incomplete piston recoil, uneven corrosion, transfer films and some other reasons. This, so-called aggressive approach, is extensively applied in Europe, mainly by German manufacturers of the FM. The major apparent drawback of this approach is a significant reduction of the disc service life.

The temperature dependencies of the mass loss from the wear per single braking were derived during the bench tests of the FM that complied to above-described compositions as presented in Fig. 7.28.

Another important property of the FM in what concerns forced vibration is compressibility or its inverse value called rigidity in compression [60]. The extent of DTV was found on the base of 2–3 factors that are dependent on the FM formula [61, 62]. In this case, compressibility should be as high as possible in order to make the contact pressure distribution most uniform [50, 63]. This helps to avoid

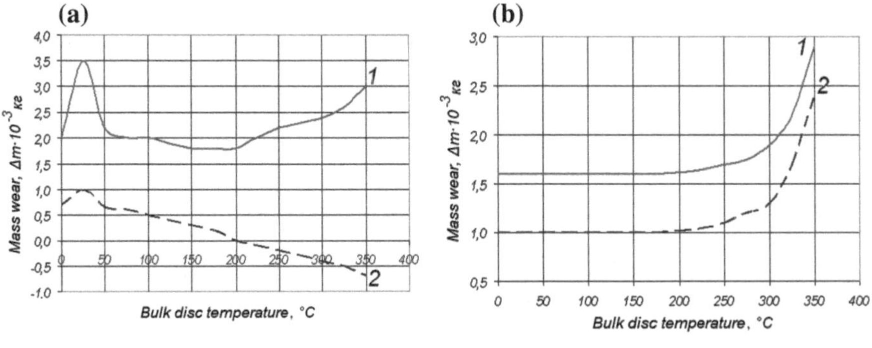


**Fig. 7.28** Temperature dependence of the mass wear of a brake disc (**a**) and friction lining (**b**) for the FM with "aggressive" (*1*) and "passive" (*2*) compositions [67]

thermoelastic instability and formation of the hot bands and thermal DTV, but raises the probability of squeal generation.

The relation between the BTV, BPV and compressibility in the modern FM is nonlinear [60]. This is in part conditioned by the fact that rigidity of a material gradually increases with increasing compressive load [64]. Therefore, a similar DTV level will promote in such materials a higher BPV level at hard braking than at smooth deceleration.

From the other hand, to ensure sufficient recoil of the piston and a more perfect pedal perception, compressibility of the FM should be the least and decreasing with increasing compressive loads. In practice, the compressibility range is rather narrow, while its deviations into one or another side may lead to reclamations [65]. The traditional composite FM [66] display elevated sensitivity (rigidity variations at compression) that increases the DTV level [5]. The materials with a linear compressibility characteristic do not require any higher safety factor in contrast to the traditional ones because of the original DTV (thermal DTV, or induced by prolonged uneven wear, and so on). In other words, the linings with a linear characteristic are applicable in the discs with above 10 μm tolerance for thickness. An alternative method to ensure a perfect pedal perception is to use a new system with fit with electric control.

The novel compromise decisions for the FM formulas are created in response to advanced scientific evidences in physico-chemical phenomena observed in frictional materials during interactions and their effect on wear behavior of the metallic counterbodies. They may, probably, eliminate above-described drawbacks of the two approaches. It is very actual today to study the effect of the FM composition upon nonuniform wear of the disc and DTV appearing at periodic contacting when the brake is disengaged [67, 68].

It has been established that the transfer films formed on the brake disc surface contain elevated amounts of barium sulfate and carbon. They display much higher wear resistance than the films containing copper or grease derivatives and metal sulfides [69–72]. No doubt, this is the frictional composite that is the source of named substances in the transfer films. This once again proves that forced vibration in brakes is an urgent problem of the modern frictional materials science.

One of the promising approaches to control structure and properties of polymer composites intended to improve their performance consists in exposure of the composites and their ingredients to the high-energy physical effects [73–75]. Numerous available in this sphere publications have visualized that there is a probability to perform structural transformations in non-ferrous metals [76, 77] and nonmetallic materials [78] by exposing them to the pulse magnetic fields. In particular, the elevated structural dispersity and plasticity of copper and its alloys, and some other effects have been disclosed. In this connection, we may anticipate that exposure to magnetic fields aimed at perfecting a series of properties may turn to be effective for the polymeric materials as well.

The probable mechanisms of the magnetic pulse treatment effect on the properties of polymeric materials are considered in [79]. This work shows on the example of such fibrous polymer materials as viscose, polyacrylonitrile, polyamide, cotton cellulose, and natural wool how the pulse magnetic field with a maximal intensity of 440 A/m influences mechanical properties of these materials.

Nevertheless, we have not reached a complete understanding of the mechanism of magnetic effects in this category of investigations, and they themselves are questionable because of the problems in reproducibility. In this connection, it is critical to achieve unbiased additional proofs to the effect of high-energy physical methods on physico-chemical properties of polymer composites.

## References


1. I.V. Kragelskii, V. Gitis, *Friction-Induced Self-Oscillations* (Nauka, Moscow, 1987), p. 184
2. V.P.Sergienko, S.N. Bukharov, A.V. Kupreev, Noise and vibration in brake systems of vehicles. Part 1: experimental procedures (review). J. Frction Wear **29**(3), 234–241 (2008)
3. W. Liu, J. Pfeifer, *Introductions to Brake Noise & Vibration, Honeywell Friction Materials* [Electronic resource]. http://www.sae.org/events/bce/honeywell-liu.pdt
4. H. Ouyang, J.E. Mottershead, Friction-induced parametric resonances in disc: effect of a negative friction-velocity relationship. J. Sound Vib. **209**(2), 251–264 (1998)
5. H. Jacobbson, Aspects of disc brake judder. Proc. Int. Mech. Eng. Part D: J. Automobile Eng. **217**, 419–430 (2003)
6. A.K. Pogosyan, V.K. Makaryan, G.S. Gagyan, Design of vibration stability of friction pairs in the disc-block brake systems of machines. J. Friction Wear **12**(2), 225–231 (1991)
7. A.K. Pogosyan, V.K. Makaryan, A.R. Yagubyan, Sound as an ecological characteristic of new frictional materials. J. Friction Wear **14**(3), 539–543 (1993)
8. M.R. North, Disc brake squeal. Proceedings Conferences on Brake of Road Vehicles, Institution of Mechanical Engineers, C38/76, pp. 169–176 (1976)
9. M. Nishiwaki et al., A Study on Friction Materials for Brake Squeal Reduction by Nanotechnology. *SAE paper* 2008-01-2581 (2008)
10. H. Jang, Compositional effects of the brake friction material on creep groan phenomena. Wear **251**, 1477–1483 (2001)
11. Y. Han, X. Tian, Y. Yin, Effects of ceramic fiber on the friction performance of automotive brake lining materials. Tribol. Trans. **51** (2008)
12. Y. Handa, T. Kato, Effects of Cu powder, $BaSO_4$ and cashew dust on the wear and friction characteristics of automotive brake pads. Tribol. Trans. **39**, 346–353
13. H. Abendroth, Worldwide brake—friction material testing standards, challenges, trends. *Proceedings of 7th International Symposium Yarofri, Friction Products and Materials*, Yaroslavl, 9–11 Sept 2008 pp. 140–150
14. H. Abendroth, B. Wernitz, The integrated test concept: Dyno-vehicle, performance-noise, B. *SAE Paper*, 2000-01-2774 (2000)
15. Bo N.J. Persson, *Sliding Friction: Physical Principals and Applications* (Springer, Berlin, 1998), p. 365
16. Yu.M. Pleskachevskii, V.P. Sergienko, Friction materials with polymeric matrix: promises in research, state of the art and market. Sci. Innov. 5(27), 47–53 (2005)
17. A. Ilintskii, Asbestos. Ind. Saf. Soc. Insur. (7), 20–22 (1997)
18. A.I. Sviridenok, S.A. Chizhik, M.I. Petrokovets, *Mechanics of the Discrete Friction Contact* (Science and Technique, Minsk, 1990), p. 272
19. A.A. Dmitrovich, G.S. Syroezhko, Sintered frictional materials, in *Powder Metallurgy and Protective Coatings in Engineering and Instrument-Making*. (Minsk, 2003) pp. 22–29
20. *Brake Noise, Vibration, and Hardness: Technology Driving Customer Satisfaction* [Electronic resource]. http://www.akebonobrakes.com

21. V. Vadari, M. Jackson, An Experimental Investigation of Disk Brake Creep-Groan in Vehicles and Brake Dynamometer Correlation. *SAE Paper*, 1999-01-3408 (1999)
22. Patent No 61-258886, Japan, 1985
23. Patent No 4702762/05, USSR, 1989
24. Patent No 2173691, Russia, 2001
25. Patent No 1460977, USSR, 1985
26. Patent No 4678818, USA, 1985
27. Patent No 94036316, Russia, 1997
28. Patent No 96113355, Russia, 1997
29. Patent No 62-149786, Japan, 1985
30. Patent No 60-203678, Japan, 1984
31. Patent No 4690960, USA, 1985
32. H. Jang, Effects of ingredients on tribological characteristics of a brake lining. An experimental case study. Wear **258**, 1682–1687 (2005)
33. S. Ganguly, K. Pastor, G. Folta, R. Lanka et al., Reduction of Groan and Grind Noise in Brake Systems. *SAE Paper*, 2011-01-2364 (2011)
34. M.G. Jacko, S.K. Rhee, *Brake Linings and Clutch Facings, Kirk-Othmer Encyclopedis of Chemical Technology*, vol. 4, 4th edn. (Wiley, New York, 1992), p. 523
35. I.G. Zedgindze, *Experimental Design for Investigation of Multicomponent Systems* (Nauka, Moscow, 1976), p. 390
36. J.A. Cornell, *Experimental Design with Mixtures: Design, Models, and the Analysis of Mixture Data*, 2nd edn. (Wiley, New York, 1990)
37. D.C. Montgomery, *Design and Analysis of Experiments*, 3rd edn. (Wiley, New York, 1991)
38. V.I. Kolesnikov, V.P. Sergienko, V.V. Zhuk, V.A. Savonchik, S.N. Bukharov, Friction joints: investigation of tribological phenomena in nonstationary processes and some optimizing solutions. *Proceedings of 7-th International Symposium on Friction Products and Materials*. Yaroslavl, 9–11 Sept 2008, pp. 25–33
39. V.P. Sergienko, S.N. Bukharov, Noise and Vibration in Frictional Joints of Machines. Tribologia **217**(1), 129–137 (2008)
40. V.P. Sergienko, N.K. Myshkin, S.N. Bukharov, O.S. Yarosh, Investigations of the effect of friction material composition on vibroacoustic activity of tribojoints, in *Proceedings of International Science Conferences Actual Problems of Tribology*. Samara (Russia), 6–8 June 2007 pp. 266–278
41. V.P. Sergienko, S.N. Bukharov, A.V. Kupreev, A study of the influence of structure of composite materials on the vibration of frictional pairs, in *Proceedings of 10th International Conferences on Tribology*. Kragujevac, Serbia, 19–21 June 2007, pp. 85–88
42. V.P. Sergienko, Frictional materials with a polymer matrix: research directions and results attained. Tribologia 5(202), 31–40 (2005)
43. V.P. Sergienko, S.N. Bukharov, Formula and structure effect of frictional materials on their damping properties and NVH performance of friction joints. *SAE Paper*, 2009-01-3016
44. V.I. Kolesnikov, V.P. Sergienko, S.N. Sychev, S.N. Bukharov, Optimization of dynamic characteristics of friction materials and their role in friction-induced noise generation. Bull. S. Cent. RAN **5**(4), 3–14 (2009)
45. V.P. Sergienko, S.N. Bukharov, Materials science approach to reduce vibration and noise in the non-stationary friction processes, in *Proceedings of 8-th International Symposium on Friction Products and Materials*. Yaroslavl, 28–30 Sept 2010, pp. 81–86
46. V.P. Sergienko, S.N. Bukharov, Vibroacoustic activity of tribopairs depending on dynamic characteristics of their materials. Mech. Mach. Mech. Mater. **9**(4), 27–33 (2009)
47. J.R. Barber, Thermoelastic instabilities in the sliding of conforming solids. Proc. Royal Soc. Ser. A **312**, 381–394 (1969)
48. T. Hodges, Development of refined friction materials, in *Proceedings of 5th International Symposium of Friction Products and Materials*, Yaroslavl, 2003, pp. 203–208
49. H. Inoue, Analysis of brake judder caused by thermal deformation of brake disc rotors, in *Proceedings of 21st FISITA Congress*, Belgrade, 1986, pp. 213–219, paper 865131

50. T.K. Kao, J.W. Richmond, M.W. Moore, The application of predictive techniques to study thermo-elastic instability of brakes. *SAE Paper*, 942087 (1994)
51. T. Steffen, R. Bruns, Hotspotsbildung bei PkwBremsscheiben. Automobiltechnische Zeitschrift **100**, 408–413 (1998)
52. S. Koetniyom, P.C. Brooks, D.C. Barton, Finite element prediction of inelastic strain accumulation in castiron brake rotors, in *Proceedings of International Conferences on Automotive Braking*. Technologies for the 21st Century, Brakes 2000, pp. 139–148
53. Thermal Judder. *Eurac technical bulletin* 00034056. [Electronic resource]—Mode of access: www.eurac-group.com/documents/thermaljudder.doc
54. S. Gassman, H.G. Engel, Excitation and transfer mechanism of brake judder. *SAE Paper*, 931880, (1993)
55. M.D. Hudson, R.L. Ruhl, Ventilated brake rotor air flow investigation. *SAE Paper*, 971033, (1997)
56. M. Donley, D. Riesland, Brake Groan Simulation for a McPherson Strut Type Suspension. *SAE Paper*, 2003-01-1627 (2003)
57. D.G. Grieve, D.C. Barton, D.A. Crolla, J.K. Buckingham, Design of a lightweight automotive brake disc using finite element and Taguchi techniques. Proc. Instn Mech. Eng. Part D: J. Automobile Eng. **212**, 245–254 (1998)
58. R. Krupka, A. Kienzle, Fiber reinforced ceramic composite for brake discs. *SAE Paper*, 2000-01-2761, (2000)
59. R.H. Martin, S. Bowron, Composite materials in transport friction applications, in *Brakes 2000, International Conference on Automotive Braking—Technologies for the 21st Century*, London, 2000, pp. 207–216
60. K. Augsburg, H. Brunner, J. Grochowicz, Untersuchungen zum Rubbelverhalten von Pkw-Schwimmsattelbremser. Automobiltechnische Zeitschrift 101 (1999)
61. R. Avilés, G. Hennequet, A. Hernández, L.I. Llorente, Low frequency vibrations in disc brakes at high car speed. Part I: experimental approach. Int. J. Veh. Des. **16**(6) 542–555 (1995)
62. A. De Vries, M. Wagner, The Brake Judder Phenomenon. *SAE Paper*, 920554 (1992)
63. P.C. Brooks, D. Barton, D.A. Crolla, A.M. Lang, D.R. Schafer, A study of disc brake judder using a fully coupled thermo-mechanical finite element model, in *Proceedings of 25th FISITA Congress*, Beijing, 1994, pp. 340–349
64. T.K. Kao, J.W. Richmond, A. Douarre, Brake disc hot spotting and thermal judder: an experimental and finite element study. Int. J. Veh. Des. **23**(3/4), 276–296 (2000)
65. B.B. Palmer, M.H. Weintraub, The role of engineered cashew particles on performance, in *Proceedings of International Conferences on Automotive Braking. Technologies for the 21st Century*, Brakes 2000, pp. 185–195
66. J.W. Richmond, T.K. Kao, M.W. Moore, The Development of Computational Analysis Techniques for Disc Brake Pad Design, in *Advances in Automotive Braking Technology*, ed. by D.C. Barton (MEP Ltd., London and Bury St. Edmunds, 1996), p. 158
67. D. Eggleston, Cold judder. *Eurac Techical Bulletin* 00029711 [Electronic resource]—1999. Mode of access: http://www.eurac-group.com. Date of access: 07 March 2012
68. S. Kim, S. Lee, B. Park, S. Rhee, A Comprehensive Study of Humidity Effects on Friction, Pad Wear, Disc Wear, DTV, Brake Noise and Physical Properties of Pads. *SAE Paper*, 2011-01-2371 (2011)
69. A. Wirth, R. Whitaker, An energy dispersive x-ray and imaging x-ray photoelectron spectroscopical study of transfer film chemistry and its influence on friction coefficient. Phys. J. D Appl. Phys. **25**, A38–A43 (1992)
70. A. Wirth, K. Stone, R. Whitaker, A study of the relationship between transfer film chemistry and friction performance in automotive braking systems. *SAE Paper* 922541 (1992)
71. A. Wirth, R. Whitaker, S. Turners, G. Fixter, X-ray photoelectron spectroscopy characterisation of third body layers formed during automotive friction braking. J. Electron Spectrosc. Relat. Phenom. **68**, 675–683 (1994)
72. A. Wirth, D. Eggleston, R. Whitaker, A fundamental tribochemical study of the third-body layer formed during automotive friction braking. Wear **179**, 75–81 (1994)

73. V.V. Klubovich (ed.), *Actual Problems in Strength* (UO VGTUU Publ., Vitebsk, 2010), p. 435
74. Yu.M Pleskachevsky, V.V. Smirnov, V.M. Makarenko, *Introduction in the radiation materials science of polymeric materials* (Nauka i Tekhnika, Minsk, 1991), p. 191
75. Yu.I Voronezhtsev, V.A. Goldade, L.S. Pinchuk, V.V. Snezhkov, *Electric and magnetic fields in the technology of polymer composites* (Nauka i Tekhnika, Minsk, 1990), p. 263
76. A.G. Anisovich, E.I. Marukovich, T.N. Abramenko, Variations in the heat state of diamagnetic metals under the effect of magnetic field. Bull. RAS Ser. Met. 6, 108–110 (2003)
77. H.G. Anisovich, Method of nonthermal changing the structure of nonferromagnetic metals and nonmetallic phases, in *Proceedings of Korea-Eurasian Seminar, Seoul*, November 2008. pp. 166–171
78. V.V. Azharonok, I.I. Filatova, I.V. Voshchula, V.A. Dlugunovich et al., Variation of optical properties of paper under the effect of magnetic component of the HF magnetic field. J. Appl. Spectro. **74**(4), 421–426 (2007)
79. A.Yu. Persidskaya, I.R. Kuzeev, V.A. Antipin, The effect of impulse magnetic field on mechanical properties of polymeric fibers. Chem. Phys. **21**(2), 90–93 (2002)

# Chapter 8
# Physiological Aspects of Human Exposure to Noise and Vibration: Normalization

The adverse affect of the elevated levels of noise and vibration on the human organism is a well-known fact. From the physical point of view, there is no principal difference between noise and vibration. The difference exists in how these phenomena are perceived by the man. Vibration is sensed by the tactile and vestibular organs, while noise by the hearing organ. Oscillations of mechanical bodies at below 20 Hz frequency are sensed as vibrations, above 20 Hz—as vibration, but with further frequency increase are perceived as a noise. Thus, under the LF oscillations (till 15 Hz) the translational vibrations are sensed by the stat iconic apparatus, while the rotational ones by the vestibular organ of the internal ear. In the contact with a solid vibrating body the vibration is perceived by the dermal nerve endings. The chapter also treats the main physiological aspects of the effect of noise and vibration on the man and the problems of their normalization.

## 8.1 Noise Affect on Human Organism

From the physiological viewpoint, noise is commonly accepted as a negative and harmful for the health factor displayed in numerous ways.

The effect of intensive noise (above 80 dB*A*) on man's ear may result in a partial or total loss of hearing. Depending on its time and intensity, man's ear is subjected to a proportional reduction of sensitivity of the hearing organ expressed in a temporary audibility threshold shear that may restore upon noise termination. If the noise lasts long and is rather intensive, an irreversible loss of hearing may occur (bradyacusia) characterized by a constantly varying audibility threshold.

In industry, we often encounter a disguised effect of the noise leading to intrusion of audibility. The degree of deadening at some enterprises is sometimes so high that it is hard to make out sound signals or speech. It is critical for a noisy production to maintain audibility such as to ensure communication between personnel and their safeness during a technological process. It is important to remember that speech obscuration influences adversely man's psyche.



V.P. Sergienko and S.N. Bukharov, *Noise and Vibration in Friction Systems*,
Springer Series in Materials Science 212, DOI 10.1007/978-3-319-11334-0_8

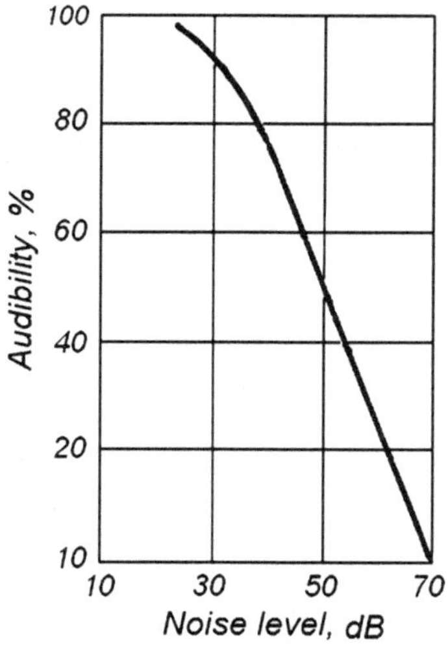


**Fig. 8.1** Audibility dependence on noise [1]

Figure 8.1 illustrates how speech discrimination depends upon noise interference. When the interference level reaches 20 dB, it does not influence speech discrimination but it impairs with increasing interference level. The discrimination level making up 75 % of the initial one (corresponds to interference level 40 dB) is considered to be satisfactory. At above 45 dB the masking noise weakens seriously speech eligibility, and beginning with 75 dB it becomes indistinct.

Although the ear is able to withstand various functional disorders for a long time in the noisy environment, accumulation of excessive irritating factors will finally hurt the hearing organ.

The aural sensation of the personnel working in a noisy environment impairs depending on the noise intensity and frequency. The minimal weighting under which the hearing organ feels weary depends upon the perceived sound frequency. Impairment of hearing can be expressed quantitatively via the shifts of the hearing threshold at different frequencies. The fatigue effect for the sounds of 2,000–4,000 Hz frequency begins at 80 dB, and at 5,000–6,000 Hz frequency it starts at 60 dB. Most often we lack the recorded audiometric data that precedes the noise effect, so injury of hearing is estimated by its thresholds. The acceptable hearing thresholds till which a man can understand a speech have been normalized.

The ear fatigue should be considered as an early warning of the loss of hearing or deafness. The syndromes of the auditory receptor disease are a headache, tinnitus, sometimes loss of balance and sickness. With the loss of hearing the eardrum undergoes thickening and slight extension together with abnormal change in the sensory endings of the auditory nerve in the spiral organ. Simultaneously, the subcortical auditory centers regulating the ear trophism are subjected to overstraining, which disturbs nutrition of the sensory cells.

Eligibility of the production noise above 80 dB*A* is estimated today from the positions of its effect on man's ear. The degree of injury of the hearing organ depends on the sound level, its duration and individual tolerance of the man.

According to medical statistics, bradyacusia has lately occupied a leading place among professional diseases and shows no tendency to any reduction [2, 3].

Not only the ear is exposed to the noise. Irritation by noise is transferred via the tissues of the auditory nerves to the central and vegetative nervous systems, affecting thereby the internals and leading to functional disorders in the organism. The psychic state is also injured, which is displayed in the anxious feeling and annoyance. Exposure of the vegetative nervous system to noise becomes evident even under low enough sound levels (40–70 dB*A*), which is independent of the subjective perception of the man. Most expressed of the vegetative reactions is the disturbance of the peripheral blood circulation as a result of constriction of the cutaneous and mucous capillaries, and boost of arterial tension (above 85 dB*A*). In contrast to the vegetative nervous system characterized by a clear compliance between the noise and response to, there is no such conformity in the human psychics. The psychic reactions are known to appear as a response to a 30 dB*A* sound. The decisive factor in psychic evaluation of the nuisance degree is individual perception. The weight on psychics augments with increasing noise level and frequency, or with the diminishing frequency bandwidth.

Exposure of the central nervous system to noise extends the latent period of the visuomotor response, disturbs mobility of the nervous processes, leads to variations in the electroencephalographic data, and violates the bioelectric cerebral activity with the apparent functional changes in the organism. The noise of 50–60 dB*A* results in essential changes in the cerebral potentials and biochemical changes in the cerebral structures.

The pulsed or irregular noise strengthens its effect. The changes in the functional state of the central and vegetative nervous systems are detected much earlier and under a lower level as compared to the auditory sensation reduction.

The "noise sickness" is characterized by a complex of medical symptoms. To the objective symptoms we refer hearing impairment, indigestion expressed in reduced acidity, cardiovascular inefficiency, and neuroendocrine disorders.

The staff working in conditions of prolonged noise nuisance experience irritability, headache, giddiness, loss of memory, elevated fatigability, loss of appetite, pain in the ears, and so on. Above-named disorders in the organs and systems of the organism may result in negative changes of the emotional state of a man till a stress situation. Noise exposure reduces attention focusing, disturbs physiological functions, leads to weariness connected with intensified metabolic costs and neuropsychic strain, and impairs speech illegibility. Above-named factors lead to a decrement of performance, along with quality and occupational safety. Figure 8.2 shows labor productivity indices as a function of the mean noise level during a working day. The noise level increased from 70 till 100 dB*A* results in a 30 % productivity decrement in the works requiring hyper-attention, which cuts production profitability. It was also proved that the total sickness rate of the workers at the noisy production sites is by 10–15 % higher. A hypothesis has been put forward

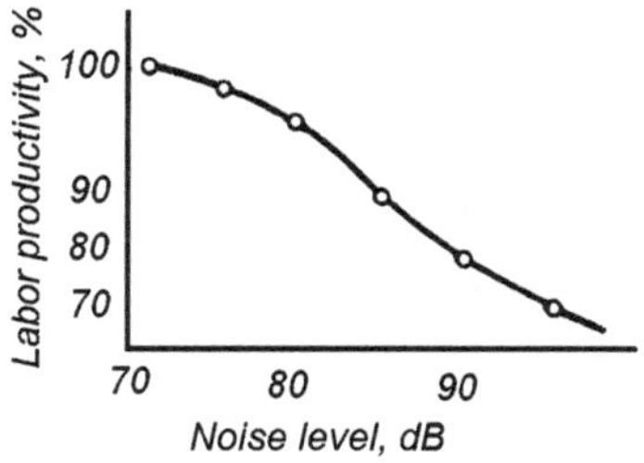


**Fig. 8.2** Labor productivity as a function of noise level

proceeding from the concept of the noise effect on the whole organism, that the average noise levels (below 80 dB*A*) harmless for hearing may, nonetheless, exercise an adverse fatiguing effect added to a similar influence of the labor weight and intensity. It was proposed to postulate the identity and synergism of the noise effect as a component of the working medium and the labor weight on the whole human organism, i.e., on the operator [4].

A concept has been put forward in [4] that the noise exposure is equivalent to the nervous load. The authors proceeded from a supposition on the primacy of the mid level noise volume effect of the on the nervous system of both direct and remote factors by taking into account that twice as much change in the volume corresponds to the sound level change by 10 dB*A*. This concept has been verified by the research based on the social and hygienic approaches, as well as physiological and clinical methods and evidences. This concept forms the base of the technical-normative and legal acts rationing noise levels on the working places with account of the labor intensity and weight [5].

## 8.2 Vibration Effect on Human Organism

Vibration belongs to the factors of a strong biological activity. The behavior, direction and depth of physiological deviances in different functional systems depend upon the level, spectral composition of vibration and physiological features of man's body. An important role in the genesis of these reactions is played by the vestibular, visual, cutaneous and other analyzers. Vibration may lead to disorders in the cardiac activity, nervous system, to vessel cramp, changes in the joints till limitation of movements. A long-lasting exposure to vibration results in obstinate pathological disorders in the organism. A profound analysis of above-considered pathological process has laid the base for relating it to an independent nosological kind of occupational disease called a vibration sickness. Its efficient curing is possible only on the early stage of the illness. Therefore, frequent occurrence of irreversible changes in the organism leading to disability is an acknowledged fact.

We should also mention an important role of biochemical features of the human body in a subjective perception of vibration. These include the influence of physical phenomena on the contact surface; propagation of oscillation in the tissues, direct response of the organs and tissues, as well as irritation of mechanoreceptors generating the reactions in neuroreceptors and subjective responses.

We have today a vast experimental and clinical evidence confirming the regulating role of the central nervous system (CNS) in the occurrence of functional shifts in the neuromuscular apparatus of the persons exposed to vibration. These investigations have proved that the disorders in the motor function induced by vibration exposure may be a result of just as disturbance of the regulatory interactions in the CNS so the direct affection of the muscles. The domination of diffusive shifts can be attributed on the main to variations in the superspinal structures, while more expressed local changes in the muscles are evidently connected with their direct injuring. The sections of the sympathetic nervous system regulating the tone of peripheral vessels and those of the peripheral nervous system connected with vibration and tactile sensibility are especially sensitive to the local vibration.

Vascular disorders depend first of all on the vibration parameters. The spastic phenomena in the capillaries take place at the vibrations above 35 Hz, below this, mainly the capillary atony or a spastico-atonic state may arise. The frequency range between 35 and 250 Hz is most hazardous in respect to the vasospasm formation.

Vibration may interfere directly the performance or affect obliquely the productivity. Some of the authors consider vibration as a stress factor affecting the psychomotor functions together with emotional and mental spheres of human activities, and augmenting the probability of emergency situations.

Depending on which part of the body is subjected to the mechanical vibration, we differentiate between the local and the whole-body vibration. When a local vibration occurs, only the body part contacting the vibrating surface experiences shaking. This may be the hands holding some tool, a vibrating object or a machine part. Sometimes, the local vibration is transferred to the parts articulated with the joints directly exposed to vibration. The vibration amplitude of these parts is, however, smaller since the oscillations transferred over the tissues, especially over the soft ones, are gradually attenuating. The whole-body vibration propagates through the whole body, originating as a rule, from the vibrating surface where a workman performs his work (the floor, seat, vibrating platform or else).

The vestibular irritants, to which vibration also belongs, disturb time perception and estimate, and reduce the information processing rate. It was proved in a series of works that the LF vibration violates coordination, and most expressed changes are observed at about 4–11 Hz frequencies.

The vibration sickness occupies one of the leading places among occupational diseases. The causes of this phenomenon lie in the extensive use of the portable power tools that contradict the sanitary norms. This is also the rapid pace of labor specialization, due to which the workmen may be subjected to a lasting vibration. The risk to undergo the vibration sickness grows with increasing vibration time and intensity, in which the individual sensitivity plays a critical role. The harmful effect of vibration is intensified by the noise, chilling, fatigue, muscular strain, alcoholic intoxication, and some other factors.

The vibration sickness resulted from the local vibration is characterized by pain in the hands, more often at nights, finger whitening in a cold weather, general indisposition, shortness of temper, and probable heart pain. The main clinical symptom of this disease is circulatory disturbance of the peripheral vessels. Initially, the vascular disturbance is observed in the arm that is more intensively subjected to vibration. Later, as the disease progresses, this process spreads not only on the vessels of the other arm but also on the feet, heart and brain. This sickness is accompanied by the pains and disturbance of sensibility in the hands and very often in the feet. Pain sensation is affected severely, and intensifies together with the temperature drop on the skin of hands and feet. The obtrusion intensifies with time and gravity of the sickness. This leads to disorders in endocrine glands, internals, and metabolic processes. The exposure to a high-amplitude vibration disturbs muscles, ligaments and bones. The patients feel weakness, fatigability, loss of temper, headaches and insomnia.

In contrast to the local vibration, exposure to a whole-body vibration may lead to the clinical symptoms connected with disorders in cerebration. This affects severely the vestibular apparatus, causing headaches and dizziness. The intensity of the pathological process is subdivided into 4 stages: I—initial, II—moderately expressed, III—expressed, IV—generalized (occurs very seldom). Along with the stages, the disease is noted by the next most typical syndromes: angiodystonic, angiospastic, vestibular, and causing vegetative polyneuritis.

The low-frequency whole-body vibration, especially the resonant one, may cause a lingering trauma of the intervertebral cartilage and bone tissues, shift in the organs of the abdominal cavity, changes in the intestinal and stomach mobility, painful sensation in the loin, may also give rise to degenerative changes in the spine, chronic lumbosacral radiculitis and chronic gastritis.

## 8.3 Normalizing of Noise and Vibration

To restrict the magnitude of noise and vibration, it is critical to develop scientifically grounded restrictions for noise and vibration characteristics, i.e., to normalize noise and vibration. We differentiate between two types of normalizing, namely, sanitary (setting standards for noise characteristics on the workplaces and recreation sites) and technical norms (limitation of tolerance noise and vibration levels generated by various machines). The sanitary normalizing limits the general noise impacting a man independently of the behavior and amount of the noise sources. The technical norms are set with account of the purpose and operation conditions of the machine. In this connection, we do not have today any unified technical norms on the noise and vibration.

In a number of cases, some sanitary and technical norms have been developed with respect to the labor activity or a kind of the machine generating noise and vibration, which differ radically from one another (e.g., [5–8]). The International Organization for Standardization (ISO) recommends using a family of limitary spectra (LS) for noise normalizing [6, 8]. These spectra take into account the curves of the equal volume (see Fig. 1.1, Chap. 1) of the hearing apparatus of a man (Fig. 8.3).

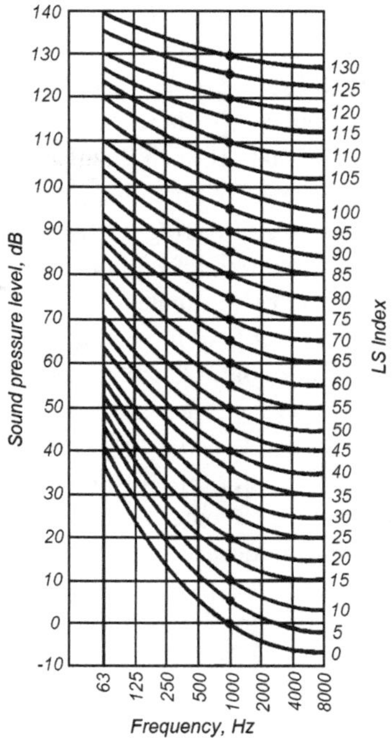


**Fig. 8.3** Forms of limitary noise spectra recommended by ISO

Normalizing with the usage of the limitary spectra includes a requirement imposed on the noise spectrum measured in dB within the standard octave bands in order not to exceed a given LS (the LS number is defined by its level in the octave band with the band center equal to 1 kHz). In some cases it is recommended to supple ment the normalizing with the use of LS by a limitation on the integral noise level defined in the dB*A* scale.

It should be noted that sometimes the branch norms on the noise are not restricted to the LS and the dB*A* levels but show rather elaborated character (e.g., the noise norms developed in civil aviation [8, 9]).

Recently, the norms on vibration in the ultrasound and infrasonic frequency range have been standardized [10].

To estimate vibration effect on the human health, comfort and sensitivity similarly to that of the noise, the whole frequency range has been subdivided into the main ranges. Vibration levels are measured not in all separate frequency ranges but only in some frequency bands (intervals) of the octave and one-third octave bands. The ratio of the octave upper the boundary frequency level to the lower one is $f_в/f_н = 2$, that of the one-third octave is $\sqrt[3]{2}$. Vibration is normalized proceeding from the parameters of vibration velocity and acceleration. Remembering that the absolute values of vibration parameters are employed within a wide range, it is convenient to use in practice the notion of the levels of vibrovelocity and vibroacceleration parameters.

The tolerance vibration levels are dependent on the frequency and kind (category) of vibration. The perception threshold for a man in the LF vibration (2–100 Hz) level corresponds to a vibroacceleration equal to 0.05–0.1 m/s$^2$. As for the vibration with accelerations 3–4 m/s$^2$ it is inadmissible [11].

The State Branch Standard GOST 12.1.012-2004 subdivides vibration according to its means of spreading in the man into a whole-body vibration that is transmitted through the bearing surfaces to the man's body, and the local vibration is transmitted mainly through the arms. Vibration operates along the axes of the orthogonal system of coordinates *XYZ*. For the whole-body vibration *Z* is a vertical axis that is perpendicular to the bearing surface; *X* is horizontal from the spine to the chest, and *Y* is horizontal from the right shoulder to the left one [12]. During the local vibration the *X*-axis coincides with the enveloping one, axis Z lies in the X plane, being directed to a feed or force application (Fig. 8.4).

At present, we distinguish six categories of the whole-body vibration. It is subdivided according to the vibration source into: 1—transport vibration generated during motion of a vehicle; 2—transport-and-technological vibration generated by the machines performing a technological process; 3—technological vibration that appears from the operating stationary technological equipment or is transmitted to the jobsites devoid of the vibration sources. The whole-body vibration of category 3 is subdivided into the following types:

(a) on the fixed workplace of production sites;
(b) on the workplace of warehouses, at the canteen, workmen's shelter, duty stations or other workrooms devoid of machines generating vibration;
(c) on the workplaces in administration offices, design bureaus, laboratories, training rooms, computer centers, health units, offices, workrooms and other places for brain workers.

The whole-body vibration is normalized for each direction of the following frequency bands: 0.8; 1.0; 1.25; 1.6; 2.0; 2.5; 3.15; 4.0; 5.0; 6.3;…80.0 Hz. The normalizing is chosen according to the intensity by taking a most intensive direction.

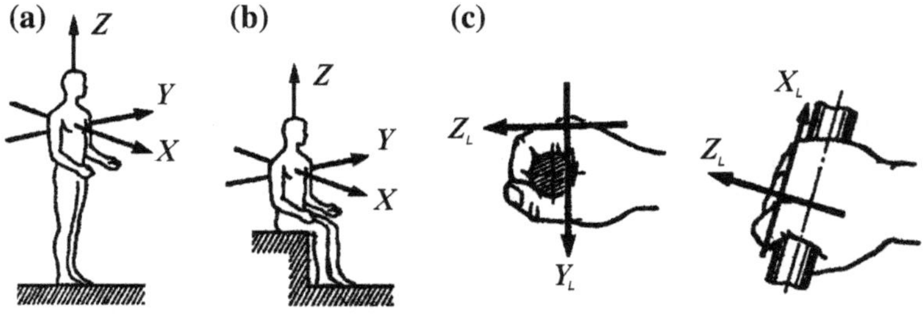


**Fig. 8.4** Direction of the coordinates for the whole-body (**a**, **b**) and local vibrations: **a**—upright position; **b**—sitting position; *Z—vertical axis* perpendicular to the surface; *X—horizontal axis* from the spine to chest; axis *Y—horizontal* from the *right* to the *left* shoulder; position of the hand under local vibration on a spherical and cylindrical surface

There also exist the norms on the local vibration influencing, e.g. the workers dealing with the mechanized hand tools. These norms are operating within a much wider frequency range in contrast to the whole-body vibration: 8; 16; 31, 5; 63; 125; 250; 500; 1,000 Hz [11].

The noise and vibration norms are revised from time to time with a tendency to their gradual toughening.

## References


1. S.P. Alekseev, A.M. Kazakov, N.N. Kolotilov, *Noise and Vibration Abatement in Mechanical Engineering* (Mashinostroenie, Moscow, 1970), p. 208
2. E. Andreeva–Galanina, S. Alekseev et al. (eds.), *Noise and Noise Disease* (Meditzina, Leningrad, 1972), p. 303
3. E.N. Ilkaeva, A.D. Volgareva, E.R. Shaihlislamova, Estimation of the probability of the occupational hearing disorder formation for the operators subjected to industrial noise. Labor Med. Ind. Ecol. **9,** 27–30 (2008)
4. G.A. Suvorov, L.N. Shkarinov, E.I. Denisov, V.G. Ovakimov, Theoretical bases of hygienic normalizing of noise. Bull. AMS USSR 1, 62–66 (1981)
5. Noise in the jobsites, motor vehicles, residential and public buildings, and on the territory of apartment blocks: Sanitary regulations and norms (SanPiN) of Nov. 16, 20111, No. 15, Minsk, Rep. Center of Hygiene, epidemiology and public health, p. 22 (2011)
6. Acoustics. Description, measurement and estimation of ambient noise. Part 1. The main units and evaluation methods. ISO/R 1996:1971; [Electronic resource] (2012), http://www.iso.org/iso/catalogue_detail?csnumber=28633. Accessed 21 March 2012
7. Acoustics. Description, measurement and estimation of ambient noise. Part 2. ISO 1996-2:2007. Determination of amvient noise levels. [Electronic resource] (2012), http://www.iso.org/iso/catalogue_detail?csnumber=41860. Accessed 21 March 2012
8. A.G. Munin, V.E. Kvitka (eds.), *Aviation Acoustics* (Mashinostroenie, Moscow, 1973), 446p
9. A.M. Mhitaryan (ed.), *Noise Reduction of Jet Airplanes* (Mashinostroenie, Moscow, 1975), p. 264
10 E.Y.Yudin (ed.), *Noise abatement in industry*. (Mashinostroenie, Moscow, 1985), p. 400
11. Industrial vibration, vibration in apartment and public buildings. SanPiN 2.2.4/2.1.8.10-33-2002. Minsk, Republican Center of Hygiene, Epidemiology and Public Health, p. 22 (2002)
12. Vibration safety. General requirements. State Standard GOST 12.1.012-90, Moscow, Standartinform, p. 20 (2004)

# Chapter 9
# Conclusions

The present book offers the analysis of investigation results on vibroacoustic phenomena in nonstationary friction processes, as well as modern tribological views on explanation of noise and vibration excitation mechanisms in the friction systems. The analytic, numeric, as well as design and experimental methods used to study brake noise are described and their disadvantages are analyzed from the standpoint of obtaining the adequate design models. The results of predicting vibroacoustic activity of friction joints by the design methods are considered.

Various design methods developed lately to simulate the low-frequency vibration are insufficiently elaborated for reaching the results commensurable to experimental ones. What concerns the approaches to simulation in the field of low-frequency vibration, have been only outlined. The design methods for the studies and prediction of the high-frequency vibroacoustic phenomena in the brake systems have become more widespread. Actuality of the design methods stems from the possibility to try-out several construction variants, to optimize them in terms of vibroacoustic parameters, and because of a high cost of the design error.

Special emphasis is placed in Chaps. 5 and 6 on the methods of damping and eliminating the sources of friction instability and related vibroacoustic activity of tribojoints. These methods can be divided in general into two large groups, namely: (i) the methods based on materials science approaches connected with perfection of frictional and viscoelastic characteristics of tribojoint materials, and (ii) the design and optimization methods of dynamic characteristics of the friction unit as a whole. The friction-induced self-vibration can be lowered till some level dependent on the friction joint design by using both groups of methods, but its compete elimination can be reached only by the former one.

In this connection, the creation of a broad class of friction materials showing a stable friction coefficient within the operating sliding velocity and temperature range, optimal thermophysical, staticokinetic and damping properties presents undoubtedly one of most promising directions in abating noise and vibration in the brake systems.



V.P. Sergienko and S.N. Bukharov, *Noise and Vibration in Friction Systems*,
Springer Series in Materials Science 212, DOI 10.1007/978-3-319-11334-0_9

The book contains summarized information on the friction materials and their vibroacoustic test procedures. The key problems in optimizing structure and composition of the friction composites based on the polymeric matrix are touched upon with the accent on surmounting noise and vibration problems in tribojoints.

Along with above-mentioned unsolved issues in frictional materials science, there remains a problem of a high-cost development stage of the friction composites with improved vibroacoustic characteristics, which is connected with a necessity to check and optimize a number of properties of these materials in a wide range of operating temperatures. For instance, during the initial design stage, the principal characteristics (friction and wear) are optimized in the laboratory and bench tests. It is only on the final stages of the bench and ride tests when a real picture can be obtained that describes susceptibility of a friction material to generating noise and vibration.

As for the theoretical analysis, it is complicated by the intricacy of the composite material formula and high sensitivity of the friction and wear properties to any composition variations. In practice, we always experience a necessity in a large number of experimental investigations in order to obtain a reliable result. Besides, there isn't any universal approach at present to selecting a composition that allows for a compromise between all characteristics of the material or makes it possible to control at least one, leaving intact all other properties.

It is a real challenge to seek for the new in principle approaches and means of changing the structure and properties of the composite materials similarly, e.g., to the high-energy physical effects on the composites or their ingredients.

The authors do not pretend to give an exhaustive disclosure of the problems concerning vibroacoustic activity of tribojoints. The book is bounded by the scope of scientific interests and is devoted to description of the methods the authors pertain to. It will be justly to anticipate some criticism of certain debatable points. So, we are looking forward with gratitude to the comments and requests of the readers on the essence of the problems touched upon in this book.

# Index





V.P. Sergienko and S.N. Bukharov, *Noise and Vibration in Friction Systems*,
Springer Series in Materials Science 212, DOI 10.1007/978-3-319-11334-0

MIX
Papier | Fördert gute Waldnutzung
FSC® C083411

Kolibri 360 GmbH
Ferdinand-Jühlke-Straße 7
99095 Erfurt, Deutschland
produktsicherheit@kolibri360.de